ELEMENTS OF QUANTITY SURVEYING

ALSO AVAILABLE

More Advanced Quantity Surveying
Christopher J. Willis & Leonard Moseley
with Don Newman

Practice and Procedure
for the Quantity Surveyor
Ninth Edition
Christopher J. Willis & Allan Ashworth

Specification Writing
for Architects and Surveyors
Ninth Edition
Christopher J. Willis & J. Andrew Willis

CESMM2 in Practice
R. G. McCaffrey, Richard G. McCaffrey
& Michael J. McCaffrey

SMM7 Reviewed
Dearle & Henderson

ELEMENTS OF QUANTITY SURVEYING

Eighth Edition

Christopher J. Willis
FRICS, FCIArb
and
Don Newman
FRICS

OXFORD
BSP PROFESSIONAL BOOKS
LONDON EDINBURGH BOSTON
MELBOURNE PARIS BERLIN VIENNA

BSP Professional Books
A division of Blackwell Scientific
Publications Ltd
Editorial offices:
Osney Mead, Oxford OX2 0EL
25 John Street, London WC1N 2BL
23 Ainslie Place, Edinburgh EH3 6AJ
3 Cambridge Center, Cambridge
MA 02142, USA
54 University Street, Carlton
Victoria 3053, Australia

First published in Great Britain 1935 by
Crosby Lockwood & Son Ltd
Second Edition 1940
Third Edition 1943
Reprinted 1944, 1945, 1946, 1947
Fourth Edition 1948, reprinted 1949, 1950,
1951, 1953, 1955, 1957, 1961, 1962
Fifth Edition 1963, reprinted 1965
Sixth Edition (metric) 1969
Reprinted 1974 and 1976 by
Crosby Lockwood Staples
Seventh Edition 1978 published by
Granada Publishing Limited in
Crosby Lockwood Staples
Reprinted 1979
Reprinted 1980 by Granada Publishing Ltd –
Technical Books Division
Reprinted 1982
Eighth Edition published by BSP Professional
Books 1988
Reprinted 1989 (twice), 1991

Set by DP Photosetting, Aylesbury, Bucks
Printed and bound in Great Britain by
Billing & Sons Ltd, Worcester

DISTRIBUTORS

Marston Book Services Ltd
PO Box 87
Oxford OX2 0DT
(*Orders:* Tel: 0865 791155
Fax: 0865 791927
Telex: 837515)

USA
Blackwell Scientific Publications, Inc.
3 Cambridge Center
Cambridge, MA 02142
(*Orders:* Tel: (800) 759–6102)

Canada
Oxford University Press
70 Wynford Drive
Don Mills
Ontario M3C 1J9
(*Orders:* Tel: (416) 441–2941)

Australia
Blackwell Scientific Publications
(Australia) Pty Ltd
54 University Street
Carlton, Victoria 3053
(*Orders:* Tel: (03) 347–0300)

British Library
Cataloguing in Publication Data

Willis, Christopher J. (Christopher James),
Elements of quantity surveying.—8th ed.
1. Great Britain. Quantity surveying
I. Title II. Newman, Don III. Willis,
Arthur J. (Arthur James),
Elements of quantity surveying
624.1′042′0941

ISBN 0–632–02219–1

Whilst every care has been taken to ensure the accuracy of the contents of this book, neither the authors nor the publishers can accept any liability for loss occasioned by the use of the information given.

Contents

Table of examples

Preface

July 1985 marked the fiftieth anniversary of the first publication of this book and it is with a mixture of satisfaction and determination that we approach the writing of this golden jubilee edition. Satisfaction that the book has stood the test of time and determination that the eighth edition should reflect the teaching of the skills of measurement and the preparation of contract documentation as practised today as the first edition did in 1935.

During those fifty years quantity surveying has changed out of all recognition. While measurement for and the preparation of bills of quantities remain an important part of the quantity surveyor's work, it is now only one of the many services that the profession has to offer. Successive editions of the Standard Method of Measurement have each reduced the number of items required to be measured and the current seventh edition is no exception. Construction, too, is generally simpler but the role of the text book remains the same, namely to explain to the student, and sometimes the practitioner, the skills of measurement and how the rules of measurement should be applied.

The latest edition of the Standard Method of Measurement, the seventh in its long history, together with the Measurement Code were published in January 1988. These rules and the code are amongst the first documents to be produced embodying the principles of Co-ordinated Project Information as represented by the Common Arrangement of Work Sections for Building Works. This is an important step forward and one which we have endeavoured to reflect both in measuring the examples in accordance with SMM7 and in the text generally.

We are sometimes criticised that the arrangement of this book has changed little over the years and that the world of quantity surveying in 1935 still shows through. It is therefore only after very careful consideration that we have decided to maintain the same basic structure of the book. A structure which assumes that the reader is coming completely fresh to the subject, something that we feel strongly has contributed in no little way to any success that this book may have had in the past, and we are satisfied that it is a structure that cannot be improved upon.

At the same time we have tried to ensure that modern principles of construction are shown and that the obsolete phraseology for which we are sometimes taken to task has been eliminated. As explained in the text, we have opted for traditional taking-off, on the principle that, having learnt a correct way to measure, adaptation for cut and shuffle, the computer or whatever other technological surprises are in store is a matter of course.

As forecast in the preface to the seventh edition, one major change is the diminution of the chapters on working–up. Whilst one sees from time to time advertisements for 'workers up' the implications of metrication, cut and shuffle and the growing use of the computer mean that this valuable member of a quantity surveyor's team is fast disappearing and the description of the work can be confined to one chapter.

As previously the book opens with some practical guidance on measuring generally and the application of mensuration. Specific examples of simple measurement for parts of the building follow and the book ends with guidance on the preparation of the contract documentation—guidance that is expanded upon and taken one step further to the final account in *More Advanced Quantity Surveying*.

The fact that Arthur Willis did not live to see the fiftieth anniversary of the publication of his famous book is of course a great sadness to us. As the book starts its second half century we can only hope that it will prove a fitting tribute to a man who did so much pioneer work in the field of writing modern quantity surveying text books.

C.J.W.
D.N.
January 1988

ACKNOWLEDGEMENTS

We are indebted to Ruth Hawkesworth who prepared the drawings, to Julie Lobb who typed large parts of our manuscript and to Peter Buttenshaw for writing the examples. Their efforts are gratefully acknowledged.

Preface to the first edition

My experience as both lecturer and examiner has impressed on me the apparent need for a textbook on the *elements* of quantity surveying, a book which, giving everything in its simplest form, will assist a student to a good grounding in first principles (in nothing more important than in quantity surveying). I have therefore kept solely in mind the requirements for preparing the bill of a normal simple domestic or small industrial building, and have refrained from introducing such allied subjects as pricing, variation accounts and professional practice.

It seems to me that a textbook on this subject should not attempt to repeat in detail the rules laid down in the official *Standard Method of Measurement*, a volume which it is now recognised must be in the hands of every student. My aim has therefore been to concentrate on method and procedure and to give practical examples, leaving the student to study concurrently the appropriate sections of the official volume, to the new (1935) edition of which references are given at the head of each chapter on taking-off. Such references will not all apply to every case, but are given as a guide to those clauses of the *Standard Method* which, having regard to the limits I have set myself, are most likely to be needed.

In preparing examples I have made a point of reproducing to one of the customary scales without any reduction, so that the student can use his scale on them—an essential to their proper study. The examples cannot comprise all possibilities in even the simplest building, and are intended only as a first step in the understanding of the general principles involved. The dimensions and abstract are reproduced from handwriting to illustrate the proper method of setting out the sheets, which cannot be conveyed in print. I had hoped to include the complete dimensions, abstract and bill of a small building to which the student could refer after studying the isolated examples, but practical limitations do not allow it be be incorporated in this volume, and such an example must therefore await a later opportunity.

In deciding the scope of this book I have generally limited myself to the current syllabus of the Intermediate Examination Part II (Quantities sub-division) of the Chartered Surveyors' Institution, except that I considered that Internal Plumbing, as one of the integral parts of every domestic building, should be dealt with. It is hoped that the book will also be found of use for the examinations of the Institute of Builders and City and Guilds of London Institute.

My thanks are due to The Architectural Press Ltd for permission to reproduce from my book, *Working Up a Bill of Quantities*, my suggested rules of order. I am also very grateful to Mr R. H. Stevens, FSI, and Mr C. E. Kenney, FSI, for kindly reading the proofs and making a number of valuable suggestions. I cannot but add a word of thanks to all those who as my principals, assistants, examinees, or students have by their ideas (and mistakes) contributed directly or indirectly to the subject-matter of the book.

July, 1935 A.J.W.

1 Introduction

THE MODERN QUANTITY SURVEYOR

As suggested in the Preface the training and knowledge of the quantity surveyor have enabled the role of the profession to be extended into new fields. These include cost advice to the client, advice on procurement and contractual methods, project management in all its varied forms and many others. Whilst the importance of these new roles cannot be emphasised enough, success in carrying them out stems from the traditional ability of the quantity surveyor to measure and value. It is upon the facet of measurement that this book concentrates.

THE PURPOSE OF A BILL OF QUANTITIES

Just as one asks before having a carpet laid what it will cost, so a prospective building owner wants to know before placing the order what the cost of his building will be. The architect can prepare drawings and a specification which define exactly what is wanted, but the builder cannot easily quote a price simply by looking at these documents. The carpet layer can estimate that so many metres of carpet and so many hours of labour will be required, both at known costs, to which would be added a certain amount to cover profit and overhead charges. In the case of buildings, each has its own characteristics and different conditions of site, and, except in the rare case of repeat orders, the builder must work out the probable cost in detail for each case. The amount of labour and material required for a building is a much more complicated assessment than that necessary for just a carpet.

METHOD OF ANALYSING COST

It is evident that if the building is split up into its constituent parts and the cost of each part can be estimated, an estimate can be compiled of the whole work. It has been found in practice that a schedule can be made setting out the quantity of each type of work in recognised units of measurement, and that estimated prices can be built up for the labour and material involved in each unit. This schedule is the bill of quantities, the prices in which can be added up to arrive at a total sum. It must not be forgotten that the bill of quantities in its normal use produces only an estimate. It is prepared and priced before the erection of the building and gives the builder's estimated cost. Such estimated cost, however, becomes under the normal building contract a definite price for which the builder agrees to carry out the work as set out in the bill. The bill must, therefore, completely represent the proposed work so that a serious discrepancy between actual and estimated costs does not arise.

ORIGIN OF THE INDEPENDENT QUANTITY SURVEYOR

Competitive tendering is one of the basic principles of most classes of business, and if competitors are given full information of the requirements of the case it should be fair to all concerned. Builders when tendering, however, found that a lot of work was involved in making detailed calculations and measurements to form the basis for a tender, and realised that by getting together and employing one person to make these calculations and measurements for them all in a particular case that a considerable saving would be made in their overhead charges. They began therefore to arrange for this to be done, each including the surveyor's fee in their estimate, and the successful competitor paying. Each competing builder was provided with a bill of quantities which each could be priced in a comparatively short space of time. Such was the origin of the independent quantity surveyor, and it was not long before the position was realised by architect and building owner. Here the building owner was paying indirectly for the quantity surveyor through the builder, whereas the surveyor could be used as a consultant if a direct appointment was made. In this way the quantity surveyor began to get the authority of the employer, and, being trusted by both sides, attained the independent position which is normally held after the signing of the contract, to value and measure work in progress and to measure and value variations to the original design. A relic of the earlier system remained in practice, now almost obsolete, of adding quantity surveyor's fees to the end of the bill of quantities to be paid by the builder after the first stage payment for building work was received.

ESSENTIALS IN A GOOD QUANTITY SURVEYOR

What then are the essentials in a good quantity surveyor? An ability to describe clearly, fully and precisely the requirements of the architect and arrange the bill of quantities so that the builder's estimator can quickly, easily and accurately arrive at the estimated cost of the work is essential. This being so, it is obviously important that the surveyor should be able to write clearly in language that will not be misunderstood, and must have a sound knowledge of building materials and construction and of customs prevailing in the industry. Moreover the surveyor must be careful and accurate in making calculations, have a systematic and orderly mind and be able to visualise the drawings and details. Cost advice requires a detailed knowledge of contractor's prices, experience of the building process and an ability to foresee the likely effect of economic trends. Dealing with contracts requires a certain amount of legal knowledge particularly and perhaps not surprisingly, the law of contract.

DIVISIONS OF BILL PREPARATION

The traditional preparation of a bill of quantities divides itself into two distinct stages.

(1) The measurement of the dimensions and the compilation of the descriptions from the drawings and specification. This process is commonly known as taking–off.
(2) The preparation of the bill. This involves the calculation of volumes, areas etc (squaring the dimensions). Traditionally this was followed by entering the descriptions and the squared dimensions on an abstract to collect similar items together and present them in a recognised bill order. From this abstract the draft bill was written. This process is commonly known as working–up.

With the virtual disappearance of the worker–up traditionally responsible for all calculations, abstracting and billing and the introduction of electronic calculators for squaring and checking dimensions, the abstracting stage has in many cases given way to a system of slip sorting known as 'cut and shuffle' whereby the original dimensions and descriptions having been copied, are cut into slips and sorted into bill order.

More recently by the use of computers, the dimensions and descriptions are coded and entered into the machine which produces a print–out of the bill of quantities, having calculated and sorted the items in the correct order.

Another development which is suitable for use on a computer is the formulation of descriptions from standard libraries but whilst the standard descriptions can be stored and used as required it is often necessary to write preliminaries, preambles and uncommon items (commonly called 'rogue' items) which are peculiar to the particular contract.

QUANTITIES AS PART OF THE CONTRACT

The bill of quantities usually forms one of the contract documents, the contract providing that the quantity of work comprised in the contract shall be that set out in the bill of quantities. In such a case the builder is expected to do and the employer to pay for neither more nor less than the quantities given, an arrangement which is fair to both parties. Thus it will be seen how important accuracy is in the preparation of the bill, and how a substantial error might lead a building owner to enter into a contract which involved a sum considerably beyond what was contemplated. If however, the bill of quantities is not part of the contract, as for example when a builder prepares an estimate from drawings and specification, the risk of errors in the quantities is taken by the builder. When the quantities are prepared by the architect or an independent surveyor it would be unfair that this risk should fall on the builder.

BUILDERS' ESTIMATES

The subject of quantity surveying is being dealt with in this book chiefly from the professional quantity surveyor's point of view, but the ability to prepare quantities is, of course, very necessary to the builder for the compilation of estimates where quantities are not supplied, which is common in small contracts. The builder's estimate in such cases may, on the face of it, look a rough affair: it is, perhaps, written in pencil with shortened descriptions, it may take short cuts in measurement which the quantity surveyor cannot, and has pricing worked out against the dimensions; there is no separate bill and no preambles to each work section describing materials and workmanship. But it must not be judged by appearances; it needs every bit as much care and system as does the quantity surveyor's bill. The essential difference is that the quantity surveyor's bill will be interpreted by a number of builders tendering in competition, and must therefore be complete in its information and a suitable basis for a contract. The builder's estimate is solely for internal pricing. If mistakes are made or short cuts are taken which lead to errors, the builder alone suffers. Sometimes it is possible to take as obvious what the quantity surveyor must describe in detail. The same general principles of measurement will apply in both cases, but the builder is free to adapt them to the firm's needs, whereas the quantity surveyor is bound by standard rules in the interests of all builder's tendering. Nevertheless, the builder's surveyor must be able to check the quantity surveyor's bill and measure variations on the basis of that bill. It is therefore essential that the builder's surveyor should understand how the bill is prepared, and there should be no difficulty in adapting this knowledge to suit the somewhat different requirements when preparing quantities for a builder's estimate.

DIFFERENCES OF CUSTOM

It must be understood that, as a good deal of the subject–matter of this book is concerned with method and procedure, suggestions made must not be taken as invariable rules. Surveyors will have in many cases their own customs and methods of setting about their work which may differ from those given here, and which may be equally good, or in their view better. The procedure advocated is put forward as being reasonable and based on practice. Every effort has been made to explain reasons for suggestions, so that they can be balanced against any alternative proposed. Furthermore, all rules must be adapted to suit any particular circumstances which may specially apply to the work in hand, as the wide range of possibilities cannot all be provided for in advance.

STANDARD METHOD OF MEASUREMENT (SMM)

The Standard Method of Measurement, 7th Edition, (SMM) and the Code of Procedure for Measurement of Building Works (MC), both agreed by the Royal Institution of Chartered Surveyors and the Building Employers Confederation, set out rules for the measurement and description of building work. The SMM is a document which provides not only a uniform basis for measuring building work but also embodies the essentials of good practice. If all bills of quantities are prepared in accordance with these rules then all parties concerned are aware of what is included and what is to be assumed. Without the use of such a set of rules the quality of bills of quantities can vary widely. The Code of Measurement Practice is a companion volume which clarifies and explains the rules contained in the SMM.

COMMON ARRANGEMENT OF WORK SECTIONS (CAWS)

This is a system devised to define an efficient and generally acceptable arrangement for specifications and bills of quantities. The main advantages are:

- easier distribution of information
- more effective reading together of documents
- greater consistency of content and description

CAWS includes 270 work sections commonly encountered in the building industry, 150 for the building fabric and 120 for services. They vary widely in their scope and nature, reflecting the extensive range of products and materials for use by contractors, sub–contractors and specialists which now exist. CAWS is arranged with items set out in three levels. The third and lowest of these levels are the work sections themselves and levels one and two are the headings under which they are grouped. The SMM being a co–ordinated document is drafted using the first and third of these levels.

METHOD OF STUDY

The student is advised in the first place to study Chapter 2 in order to grasp thoroughly the form in which dimensions are usually written. The student is assumed to have knowledge of elementary building construction and simple mensuration and trigonometry, but if weak in these subjects they should be studied before proceeding further with measurement. Chapter 4 gives some information on, and examples of, the practical application of mensuration most commonly met with in quantity surveying. In Chapter 5 are collected a number of notes on general procedure which should be read before attempting to study actual examples of measurement, but to which it may be found useful to turn again after having made some study of taking–off. Chapters 6 to 18 represent the clear–cut sections into which the taking–off of a small building might be divided, and these should be taken one at a time. The principal clauses of the SMM and the MC applicable are referred to in each chapter and should be studied concurrently. After the chapter has been read the examples should be worked through. The student should be able to follow every measurement by reference to the drawing. Chapter 3, relating to cut and shuffle, may be studied after the measurement examples. If, however, the reader feels the necessity to become familiar with this method of entering dimensions at an early stage it should be read before the following chapters.

The examples of taking off in this book are all isolated small parts of what could be the dimensions of a complete building and are not a connected series. When they have been mastered in their isolation it will be much easier to see how they are expanded and fitted together to make up the dimensions of a complete building. The processes in preparing the bill have been outlined, and Chapter 19 introduces a number of practical points for consideration in dealing with a complete building.

Chapters 20 to 24 deal with bill preparation, which is more logically dealt with after the taking–off as this is a subsequent process. The student may, however, study these chapters simultaneously with those on taking–off.

EXAMPLES

As explained the examples are small detached parts of dimensions to illustrate the methods of measurement of a small unit of a building. They assume that full specification clauses would be set out in preambles to the bill (see Chapter 22), and that clauses would be inserted, such as those required by the SMM for keeping excavation free from water, testing drainage etc. In line with current drawing office practice the dimensions of the drawings are figured in millimetres.

The dimensions set down in the dimension column when taking–off are given to the nearest two places of decimals of a metre in accordance with the guidance given by the SMM (GR3.2): side casts when used to calculate these dimensions are given in millimetres to ensure accuracy.

TAKING-OFF

The examples in the chapters on taking–off are written on traditional dimension paper. It is recognised, however, that the majority of taking–off today, when not coded for a computer, will be on some form of cut and shuffle paper and that the bill will be typed direct from the descriptions. The use of abbreviations, except where repeating a description already written, is little practised, as this would entail someone writing them out in full before typing. Thus, although the examples are written on traditional paper, this being considered the best system for a text book and usually what the candidate will be faced with in the examination room, they are phrased as if they were on cut and shuffle paper, abbreviations only being used for deductions where a description sufficient to recognise clearly an item is all that is required. For those who wish to use abbreviations in descriptions, a list of those commonly in use is given in Appendix 1.

2 Setting down dimensions

TRADITIONAL DIMENSION PAPER

The dimensions are measured from the drawings by the taker–off, who uses paper ruled thus:

1	2	3	4	1	2	3	4

The columns (not of course normally numbered) have been numbered here for identification. Column 1 is called the timesing column, and its use will be described later. Column 2 is the dimension column, in which the measurements are set down as taken from the drawings. Column 3 is the squaring column, in which are set out the calculated volumes, areas, etc., of the measurements in column 2. Column 4 is the description column, in which is written the description of the work to which the dimensions apply, and on the extreme right–hand side of which (known as 'waste') preliminary calculations and collections are made. There are two sets of columns in the width of a single A4 sheet. No written work should be carried across the central vertical division. There is usually a narrow binding margin (not shown above) on the left of the sheets. Sizes of dimension sheets widths of columns etc. are covered by a standard specification (BS 3327 – Stationery for quantity surveying).

Some surveyors use dimension paper in A4 sheets, others in folded A3 sheets, the choice really depending on the method of tying up or binding. Some prefer single sheets ruled on one side only. If the sheets are to be bound, double sheets should be used and all writing kept out of the binding margin.

Each dimension sheet should have the name and/or the number of the job written on or, better, stamped on, a stamp being easily made up from an office printing outfit or specially made commercially at no great cost. In addition the title of the section being measured should be written followed by a number, starting at one for each section. The examples measured in the following chapters have been written using only one half of the sheet, the right-hand side being used for explanation.

CUT AND SHUFFLE DIMENSION PAPER

An example of a sheet of cut and shuffle dimension paper and a full description of how it is used is included in Chapter 3 where the principle of slip sorting is described.

FORM OF DIMENSIONS

Before going any further it is necessary to understand the dimensions as set down by the taker–off. All dimensions are in one of five forms:

(1) cubic measurements
(2) square or superficial measurements
(3) linear measurements
(4) 'numbers' or enumerated items
(5) items

These are expressed in the first three cases by setting down the measurements immediately under each other in the dimension column, each separate item being divided from the next by a line, e.g.:

	Dimensions			Explanation
	3.00 2.00 4.00			indicating a cubic measurement 3.00 m long, 2.00 m wide and 4.00 m deep.
	3.00 2.00			indicating a superficial measurement 3.00 m long and 2.00 m wide.
	3.00			indicating a linear measurement of 3.00 m.

An item to be enumerated is usually indicated in one of the following ways:

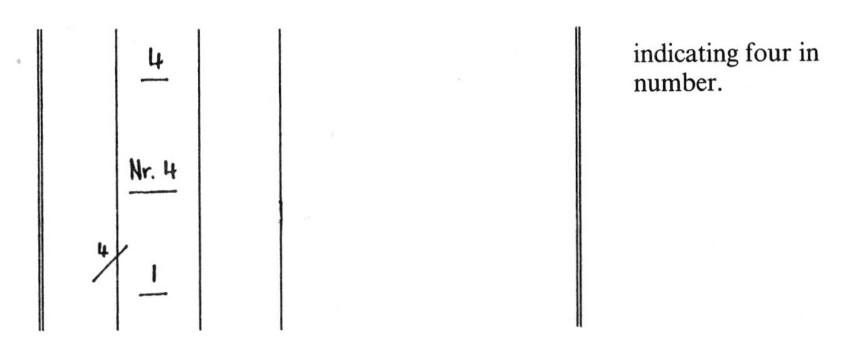

Occasionally the SMM requires the insertion of an 'item', this is a description without a measured quantity e.g. testing the drainage system. The description may, if applicable, contain dimensions e.g. temporary screens. This requirement is indicated as follows:

There is no need to label dimensions cube, super, lin, etc., as, if a rule is made always to draw a line under each measurement, it is obvious from the number of entries in the measurement under which category it comes.

It is usual to set down the dimensions in the following order:

(1) horizontal length
(2) horizontal width or breadth
(3) vertical depth or height

Although the order will not affect the calculations of the cubic or square measurement, it is very valuable in tracing measurements later if a consistent order is maintained, and, as will be explained, an incorrect order in a description may even sometimes mislead an estimator in pricing.

'TIMESING'

It often happens that when the taker–off has written the dimension it is found that there are several similar items having the same measurements, and to indicate that the measurement is to be multiplied it will be 'timesed' thus:

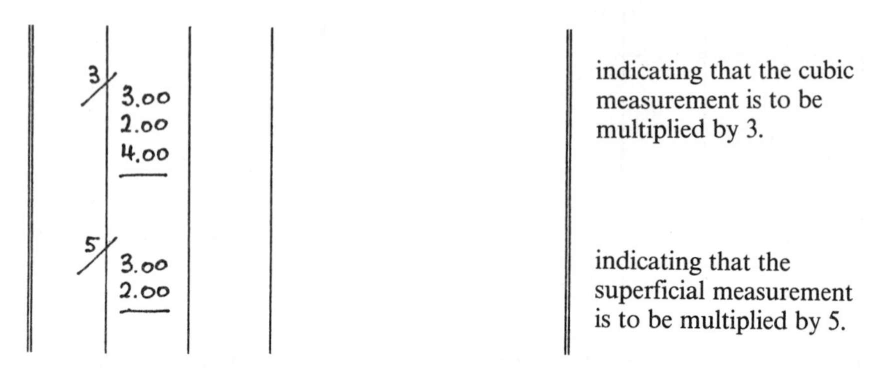

The 'timesing' figure is kept in the first column and separated from the dimension by a diagonal stroke. An item 'timesed' can be 'timesed' again, each multiplier multiplying everything to the right of it, thus:

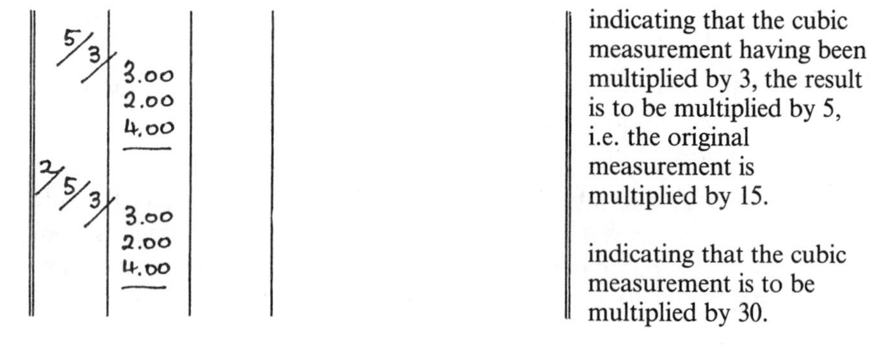

The 'timesing' is done to a linear or enumerated item in just the same way as shown above.

DOTTING ON

In repeating a dimension the taker–off may find that it cannot be multiplied but added. For instance, having measured three items as follows:

3/
3.00
2.00
4.00

suppose there are two more. The dimension could by multiplied by $1\frac{2}{3}$ thus:

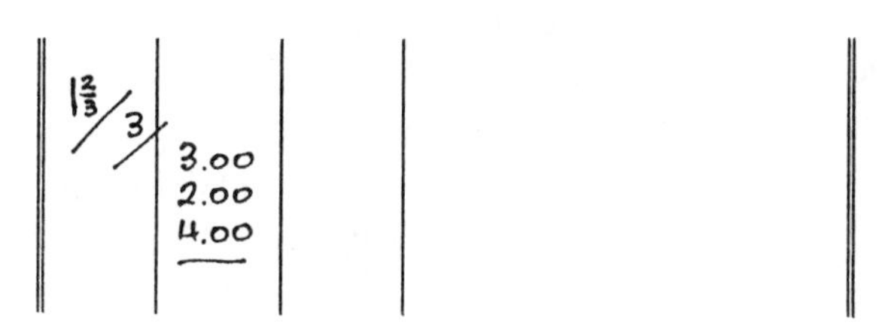

but to avoid fractions and make the train of thought clearer what is called 'dotting on' can be used thus:

2 · 3/	3.00 2.00 4.00			indicating that the cubic measurement is to be multiplied by 3 + 2, i.e. by 5.

The dot is placed below the top figure to avoid any possible confusion with decimals although these are usually avoided in 'timesing'. Figures dotted on should be lower than the last just as each one 'timesed' is usually higher, more space thus being available than if they were all written in a horizontal line.

'Timesing' and 'dotting on' can be combined, thus:

2/ 2 · 3/	3.00 2.00 4.00			indicating that the cubic measurement is to be multiplied by (3 + 2) × 2, i.e. by 10.
3 · 2/3/	3.00 2.00 4.00			indicating that the cubic measurement is to be multiplied by 3 × (2 + 3), i.e. 15.

or the last could be 'timesed', thus:

IRREGULAR FIGURES

So far only regular figures have been dealt with, but the taker–off often has to set down dimensions for triangles, circles and other such figures. Some examples of the most common figures are given below:

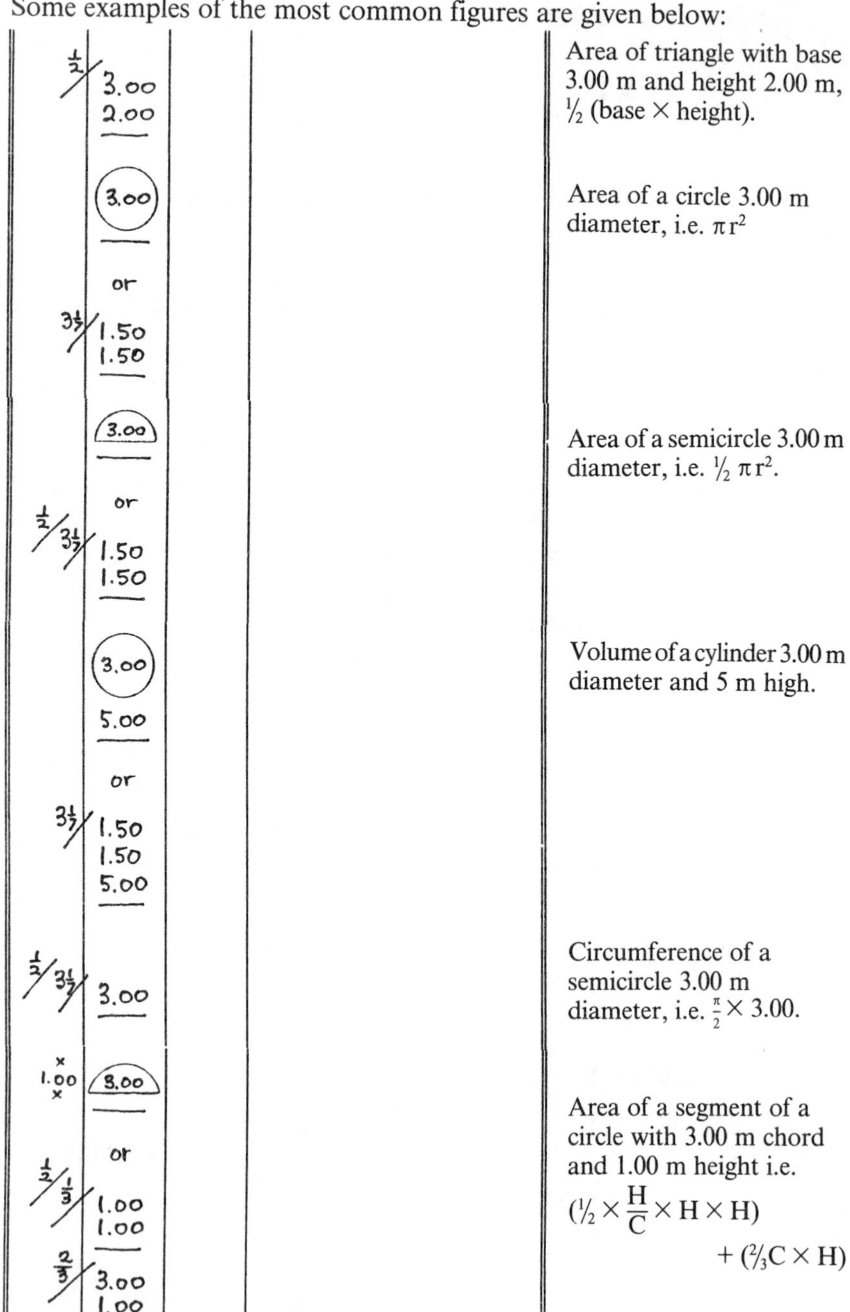

Other irregular figures will probably be measured by building up in triangles or portions of circles and one can apply a knowledge of mensuration to enter the appropriate dimensions. It is important however to ensure that 'cube' 'square' or 'linear' are readily identified from the entries. Care must be taken when writing fractions in the timesing column that the line dividing the numerator from the denominator is horizontal thus avoiding any confusion with timesing.

WASTE CALCULATIONS

Except in very simple cases, dimensions should not be calculated mentally. Not only will the risk of error be reduced if the calculations are written down, because they will be checked, but another person can readily see the origin of the dimension. These preliminary calculations, known as 'waste calculations' are made on the right-hand side of the description column. They must be written definitely and clearly, and not scribbled as if they were a sum worked out on scrap paper. The term 'waste' used for this part of the column might be thought to imply 'useless' but in fact implies 'a means to an end'. Every effort must be made to commit to writing the train of thought of the taker-off. Waste calculations should be limited to those necessary for the clear setting down of the dimensions by the taker-off, and not take the place of squaring, which is properly left to a junior member of staff. An example of waste calculations is given later in this chapter.

ALTERATIONS IN DIMENSIONS

Where a dimension has been set down incorrectly and is to be altered it should either be neatly crossed out and the new dimension written in, or the word 'nil' should be written against it in the squaring column to indicate that it is cancelled. Where a number of measurements in the dimension column are bracketed together care must be taken to indicate clearly how far the 'nil' applies, and this may be done thus:

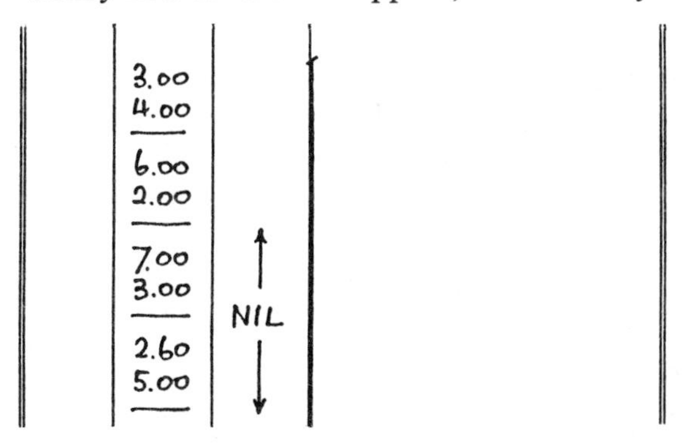

No attempt should be made to alter figures, e.g. a 2 into a 3 or a 3 into an 8. The figure may appear to have been altered satisfactorily, but may look quite different to another person or when the dimensions are photocopied. Every figure must be absolutely clear, a page a little untidy but with unmistakable figures being far preferable to one where the figures give rise to uncertainties and consequent error. Erasures with a penknife, rubber or liquid paper correcting fluid should never be made as it is often of value to know what was written in the first instance, and furthermore such an erasure is at the best of times obvious. It is best to nil entirely and write out again any dimensions which are getting too confused by alterations, but great care is needed in copying dimensions, or mistakes may be made, it being particularly easy to miss copying the timesing. Therefore where any dimensions are rewritten they should be checked very carefully with the originals.

THE DESCRIPTIONS

The description of the item measured is set down opposite the measurement in the description column, thus (the 'waste' calculation being also shown):

			34.000 16.000 50.000 2/ = 100.000 4/.215 .860 99.140
			15.050 12.000 13.500 11.700 4) 52.250 av. = 13.063 Bott. of conc. 12.050 1.013
	99.14 1.00 1.01		Excavating trenches width exceeding 0.30m, maximum depth not exceeding 2.00m

Some surveyors prefer to write the item description first and put down their 'waste' calculations below it. The writers prefer to make their calculations first and think out the description afterwards.

If two or more measurements are related to one description they are bracketed, thus:

	99.14 1.00 0.40		In-situ Concrete (21 N/mm² – 20mm aggregate) foundations poured on or against earth
	3.05 0.76 0.45		

the bracket being placed on the outside of the squaring column. As the worked–out results of each measurement have to be collected in groups according to the brackets, it is essential that a clear indication should be made to show where the bracket ends, the bracket in fact usually being a vertical line with a little cross mark to indicate top and bottom.

Where two or more descriptions are to be applied to one measurement they are written thus:

	99.14 1.00 0.60		Concrete in foundations a.b. & Deduct Filling to excavations & Add Disposal of excavated materials off site

each description in this case being separated by '&' on a line by itself. The isolated '&' always indicates a separate item in the descriptions. No bracket is necessary in this case. Care must be taken when coupling in this way a superficial with a linear item. It sometimes happens that it is very convenient to do so, but the distinction must be made quite clear so that the linear quantity is not used instead of the superficial. For example:

	5.33	5.33	25 × 100mm Wt. s.w. rounded skirting plugged
	5.79	5.79	
2/	7.00	14.00	&
	4.50	4.50	Deduct Emuls. g.s. intly
		29.62	× 0.10 = 2.96m²

To measure the deduction of the emulsion paint in this way saves setting down all the dimensions again as superficial items (exactly the same lengths being used). Similarly a superficial item might be marked to be multiplied by a third dimension to make a cube, e.g.:

	6.00 4.50		Compacting bottoms of excavations & In-situ concrete (mix) bed thickness not exceeding 150mm × 0.15 = m³

DEDUCTIONS

Where a deduction is to be made, the description is preceded by *Deduct* which is often abbreviated to *Ddt* but when carelessly written has been known to be mistaken for *Add.* It is of course important to make quite clear whether a measurement is to be added or deducted and some surveyors put the word 'Add' always to any description immediately following a deduction, others only when a following addition is coupled to the deduction by '&', as in the example given below. In taking–off on traditional paper it is important that all deductions in a series of coupled descriptions are clearly marked Deduct, all doubt whether any description is an 'add' or 'deduct' thus being removed. To add emphasis to the words Deduct and Add used in this context they are often underlined. With cut and shuffle taking–off it is necessary to keep deductions on separate slips and the fact that they are deductions being clearly indicated as described in Chapter 3.

An example of deductions follows:

	1.00 2.10		Deduct 1½ B Wall facework one side & Deduct Plaster brick walls & Deduct Emuls. plaster g.s. intly
	1.00 1.20		Deduct Gloss paint plaster g.s. intly & Add Two coats emuls. plaster g.s. intly

It sometimes happens that after such an adjustment as the last it is required to reverse the deduction and addition. Supposing the walls of a room have been measured with others as emulsion painted and deductions made accordingly, and it is required to adjust them in this one room to gloss paint, the gross area of walls will be taken as:

	15.00 3.00		Deduct Emuls. plaster g.s. intly & Add 3 Coats gloss paint plaster g.s. intly

The areas of openings in these walls will then be taken and simply described as:

4/	1.00 1.50		Less.
	1.00 2.20		

and reveals to these openings then added, as:

4/	0.25 4.00		More.
	0.13 5.00		

when 'Less' will mean 'deduct paint and add emulsion', and 'More' will mean 'Deduct emulsion and add paint'. The totals of each group will be collected so that one net measurement will be arrived at of 'deduct emulsion' and 'add paint'. Some surveyors add the words 'both last' to the 'less' or 'more' but this is considered unnecessary as it is the dimensions relating to the previous descriptions that are being adjusted rather than the descriptions themselves.

SPACING OF DIMENSIONS

One of the commonest faults found with beginners taking–off is the crowded state of their dimensions. All measurements and descriptions should be spaced well apart, so that it is quite clear where one begins and the other ends. It is not unusual for a taker–off to realise that after writing down the measurements that some item has been overlooked and it is desired to insert it in its proper place. If the dimensions are well spaced out, it can be squeezed in, but otherwise it will have to be inserted elsewhere and cross–references made which only complicate the work. The use of a few extra sheets of paper will be found well worth while.

CUT AND SHUFFLE

The same rules apply for setting down dimensions in a cut and shuffle system of taking–off as for the traditional system, except that each description and deducts have to be on new slips. An example of cut and shuffle entries are shown in the following chapter.

Timesing	Dimension	Squaring	Description
2/	28.21		Earthwork support, maximum depth not exceeding 1.00m and distance between opposing faces not exceeding 2.00m
	0.75		
	31.01		Ext. face wl. 29.420
	0.15		4/2/198¾ · 1.590
			31.010
			℄ 28.21
			131 A 45

Timesing	Dimension	Squaring	Description
	28.21		In-situ concrete (20 N/mm^2 - 20mm) foundations poured against earth
	0.70		
	0.20		
			131 A 46

Timesing	Dimension	Squaring	Description
	&		Ddt. Filling to excavations average thickness exceeding 0.25m with hardcore a.b.d.
	C	Ddt.	131 A 47

3 Cut and shuffle

When using the cut and shuffle system the taking–off is done in exactly the same way as demonstrated in this book, except that each description with its dimensions is written on a separate slip of paper. When the measurement is complete the slips are sorted into bill order and the slips containing the same descriptions are brought together and totalled on one slip. Provided that care is taken in framing descriptions and with some editing the bill can be typed straight from the sorted slips. If the word shuffle is taken to mean 'to mix at random' then cut and shuffle is perhaps a misnomer and could be better called 'cut and sort' or as sometimes referred to, 'a slip sorting system.'

It is generally accepted that this system is more economical than the traditional abstract and bill method. Advantages are that there is no repetitive writing out of descriptions and quantities and the bill can be typed soon after the taking–off is complete. On the other hand the taking–off undoubtedly takes longer. Several systems of cut and shuffle are in use which, though differing little in principle, differ in detail. Most of the remarks contained in this book relating to the procedure in taking–off a complete building equally apply when using the cut and shuffle system. The description of one such system follows.

PAPER AND RULINGS

A sheet of foolscap or A4 size paper can be divided into three sections as shown below which can be separated later by perforations or by cutting with a guillotine to form slips. The important thing is that the taker–off should do his writing on a reasonably sized piece of paper, though for sorting purposes smaller pieces are more easily handled.

Obviously if every item is to be on a separate slip many slips will have only one or two dimensions on each. There will, however, be cases where there are so many dimensions for an item that they will need two or more slips. The first slip used for an item is called the 'master'. This is the one from which the bill will be written or typed. Additonal slips for the same item, are known as 'slaves'. The first and second divisions in the illustration on page 22 are masters, the third division is a slave (its master being previously measured). It will be noted that the collection of master and all the slaves (both adds and deducts) is required to arrive at the final quantity.

The normal dimension paper ruling needs some modification if the slips are to serve as a draft bill. The item number, description and calculated quantity must be together in a prominent place where the typist can find them without confusion with other matter on the sheet. Several 'boxes' are therefore provided (see illustration below) and they are numbered for reference as follows:

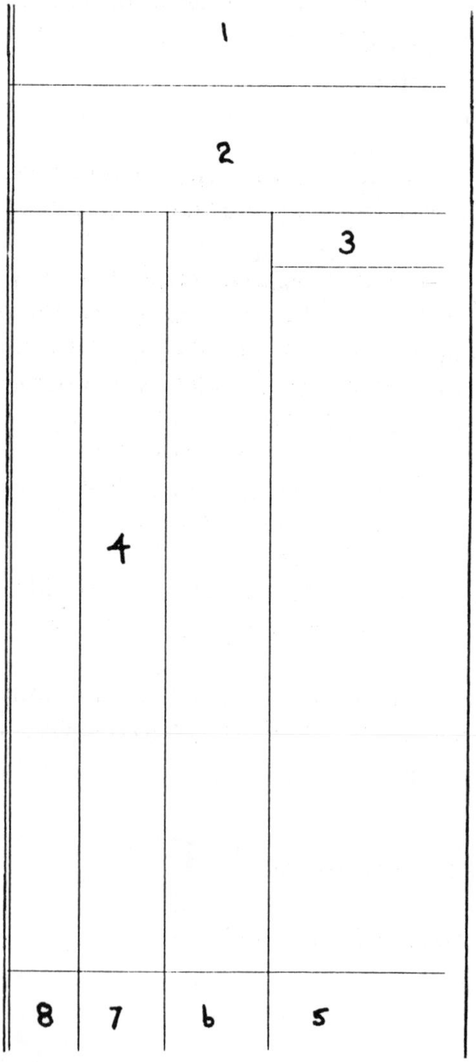

1

2

3

4

8 7 6 5

(1) Is for the bill item number if the quantity surveyor wishes to number the bill items serially. These numbers will be entered immediately before typing the bill. If referencing of items is by serial letters on each page, this box will be left blank.
(2) This space is for the description which must be written carefully and in full by the taker–off on master slips, remembering that in doing so the final draft bill is being written.
(3) When calculations are complete, this space on the master slip will have the quantity to be inserted in the bill with its unit of measurement: it will be left blank on slave slips.

These three boxes will be the only ones to be looked at by the typist and only the material to be entered in the bill must be put in them. If the description is a long one it can overflow into the top of the three dimension columns, but not into box 3.

The remaining numbers are:

(4) The normal dimension paper ruling but the right–hand column being used only for waste calculations, location notes and location headings.
(5) Reference to the job number, taking–off section and slip number, e.g. 131/A/45 (job 131, substructure, slip no.). A suggested table of taking–off sections with their references is given at the end of this chapter. The numbering of each taking–off section should start at 1, the slips being prenumbered or numbered as completed.
(6) Used for the total of the squared dimensions on the slip.
(7) In this box should be put C, S or L (cube, square or linear), and need only be used when as shown in the illustration on page 22, the category is not obvious from the dimensions.
(8) This box can be used either for the work section letter as used in the SMM or for elemental references if preparing a bill of this kind.

Note: each slip is holed in the top left hand corner so that after cutting they can be filed in batches as convenient with long treasury tags enabling the slips to be turned over easily.

RULES FOR TAKING–OFF

There are some rules that the taker–off must follow:

(1) Descriptions must be written in full as in a bill with no abbreviations, except as in 5 and 6 below.

(2) Deductions must be on a separate slip and 'Deduct' must be written in box 2 and repeated in box 6 to appear against the total. This insertion in box 6 is very important, and after completion of a section the taker–off, when looking through the dimensions, must check that all 'Deducts' are so inserted. Otherwise, the machine operator doing the casting and who is only required to look at this bottom line may make a serious mistake.

(3) Two descriptions can be coupled to a single set of dimensions by '&' by writing the second description in full on the next slip and by putting '&' in the dimension column. This is a case where C, S or L must be put in box 7.

(4) Cube items should be set down as cubes, not linear to be cubed up later.

(5) The word 'ditto' should never be used in descriptions, since when the sheets are cut there would be no indication of what is referred to. Instead on slave slips a reference should be given to the master slip number, e.g. '25 x 10 mm s.w. skirting a.b. as 1/67' and abbreviations are then allowable. The description must be full enough to leave no doubt when sorting the separated dimensions: 'skirting a.b.' is not enough, as the sorter who would not have 1/67 to hand might think that it was tile, granolithic or something else.

(6) As many slave slips as necessary for each item may be used. Normal abbreviated descriptions can be used for such slips, provided that the item is clearly identified, remembering that the slips will be separated after cutting. Slave slips will eventually be pinned to the master slip in such a way that the typist can turn over the whole batch, as the bill entry is made only from the master slip. Some surveyors prefer to detach the slave slips leaving for the typist only those which have to be copied.

(7) If a whole slip is nilled, 'nil' should be written in the item number box (1) and the quantity box (3) as well as against the dimensions. It could also be written across any description.

(8) Slips which are left blank, e.g. at the end of a section to leave space for possible later additions, must have their serial number in box 5. They will be set aside in sorting.

(9) Specialists' work measured for a basis to obtain a p.c. sum should be measured together, so that these sheets can be taken out, cut and sorted in advance of the main bill.

(10) Items which are to be grouped together in the bill (i.e. those that would have been billed direct under the traditional method) could be marked with a distinguishing letter, perhaps the taking-off section letter and another. For instance, when measuring drains all manhole dimensions could be marked XA, then, perhaps, drainage from last manhole to sewer XB. A list of the letters used and their meaning should be made. A special set of rubber stamp letters will be required for this purpose. The sorting of these slips is the equivalent of the former billing direct.

Taking-off will proceed in the normal way apart from the restriction of a new slip for each new description. Dimension headings can be written as usual across the dimension columns. It will be an advantage to have standardised section and letter references for sections of the taking-off as below, and the taker-off must see that before parting with the dimensions they are marked in box 5 as described above.

It will be found that good use can be made of schedules instead of making long collections on waste. These can be set out on A3 paper and should, of course, be preserved as a record of measurements which will be required for adjustment of variations. Such schedules should indicate clearly by room number or otherwise the position of all measurements. They should be numbered serially so that they can be referred to in side notes on the dimension sheets.

SQUARING AND CASTING

Squaring and casting are invariably done with a calculator and checked by another. The most useful type of calculator is one which gives a 'print-out' and keeps a running total which is transferred to the paper when wanted. Squaring and casting and their checking can be done on the whole sheets before cutting, but in some offices all calculations are left to be done in one operation after the editing referred to below. If all is so done in one operation, the final answer only will appear on the master slip in box 3.

If the calculator is not one which can deal with deductions in the running cast, the total of squaring will appear on every sheet. The cast of the adds will then be made and entered on the master sheet and the cast of deducts entered below it, the deduction will be made and the result usually in metres entered in box 3. This method might also be adopted because staff are available to do the squaring at the earlier stage and so get the work forward. With this procedure space must be left at the bottom of the waste column of box 4 on the master slip for the calculation to be entered.

There is normally no need for the squaring to be entered against each measurement if done by a calculator that will make a running cast. Such figures are only of value for adjustment of variations and could quite well be worked out separately if and when required.

DUPLICATE DIMENSIONS

An uncut duplicate copy of the taking–off is almost essential for future reference. A copy may be made by using NCR (no carbon required) paper or a photocopy of each sheet may be taken.

CUTTING AND SHUFFLING

Each sheet is copied before any cutting is done and the sheets to be cut must be listed, so that there is a record of the number of slips under each taking–off section. The sheets will then be separated on the perforations, or cut by guillotine, and sorted into pigeon holes, each of which represents a section of the final bill. For this a special fitting will be required with, say 36 pigeon holes made to suit the size of the cut slips. Wasted slips whether blanks or nilled will be filed together and not destroyed. By using the letter reference in Box 8 this sortation may be carried out by a person without technical knowledge.

The contents of each pigeon hole are then taken out in turn and sorted into bill order. Slave slips will be collected and pinned to their master slips (deduction slips being put at the end) and each bill section threaded in order on a treasury tag. The slips are then ready for editing. A check should be made at this stage by listing the number of cut slips in each section of the bill, adding the wasted slips and checking the total with the number of slips in the original uncut copy.

EDITING

Although under this system the abstract is saved, the work of editing is probably heavier than in the traditional method where much co-ordination was done by the biller. The editor must check the descriptions and order of items, amend descriptions to 'ditto' where applicable and insert group headings. Headings are written on blank pieces of paper, preferably coloured, of the same size as the cut slips and inserted in the appropriate places. The bill section headings and preambles may be written on normal bill paper and items to be written short are given a distinctive mark. The editor will do all the work in red or other distinctive colour and will insert in box 3 'm^3, m^2, m or Nr.', as required to be printed in the bill, leaving room for the quantity to be inserted later. A special set of rubber stamps is useful for this purpose. Notes on writing short, the use of ditto and headings are given in Chapter 21.

If extra items are found necessary in editing, they should be written on a spare blank slip, also on the blank copy with the same reference number. Similarly if a sheet is nilled, it should be nilled on the photo copy. Reference to the later chapters on preparing and editing the bill will provide considerable information on this part of the work.

PC and provisional sums will appear on dimension sheets in the normal way, but the editor may like to write the bill of these items in the traditional way, together with the preliminary bill, preambles and summary. The writing of these sections is discussed in Chapter 22. A standard set of items in a word-processor memory may be useful for producing these sections of the bill. A works on site or spot item bill will usually be written out.

REDUCING QUANTITIES AND NUMBERING ITEMS

The editor has then finished his work and the bill is ready for the typist, except for the insertion of the quantities and item numbers. The collection of totals on each master and its slave sheets is best done on a calculator which gives cumulative totals, the result being entered in the lower part of the waste column of the master sheet. The total in metres will then be written against the unit of measurement and entered in box 3 of the master slip. Both processes must, of course, be checked. Finally items will be serially numbered through the bill if that method of numbering is adopted. Some surveyors prefer a method of using letters serially from A

JOB: Southtown School.

JOB Nr: 131.

PROGRESS

DIMS COLUMN NRS.	T.C. INITIAL	TAKING-OFF SECTIONS	SQRD	CHKD	DIMS COPIED	NR. SLIPS	PHOTO COPIES CHKD	CUT
A1 - 144	KGSS	Substructure	✓	✓	✓	144	✓	✓
C1 - 162	PGB	Brickwork & Facings	✓	✓	✓	162	✓	✓
E1 - 81	PGB	Roof Const. Coverings	✓	✓	✓	81	✓	✓
I1 - 261	OW	Internal Finishings	✓	✓	✓	261	✓	✓
J1 - 117	OW	Windows	✓	✓	✓	117	✓	✓
K1 - 81	OW	Internal Doors	✓	✓	✓	81	✓	✓
L1 - 99	OW	External Doors	✓	✓	✓	99	✓	✓
M1 - 128	OW	Fittings	✓	✓	✓	128	✓	✓
Q1 - 87	CJW	Builders Work	✓	✓	✓	87	✓	✓
S1 - 201	CJW	Plumbing	✓	✓	✓	201	✓	✓
Y1 - 96	KGSS	External Works	✓	✓	✓	96	✓	✓
X1 - 111	CJW	Drainage	✓	✓	✓	111	✓	✓
Z1 - 84	CJW	Existing Building	✓	✓	✓	84	✓	✓
						1652		

SCHEDULE

DATE: October 1987.

SHEET Nr: 1.

INITIAL SHUFFLE	NR.	BILL SECTIONS	CAWS	FINAL SHUFFLE	EDIT	QUANTS CALC	QUANTS CHKD	CUT SHEETS	TO TYPE	NOTES
	1.	Preliminaries and general conditions	A	–	✓	–	–	–	✓	Billed Direct.
✓	–	Complete buildings	B	–	–	–	–	–	–	Not applicable.
✓	2.	P.C. and Provisional Sums		–	✓	–	–	20	✓	
✓	3.	Demolitions, Alterations and Renovation	C	✓	✓	✓	✓	–	✓	Billed Direct.
✓	4.	Groundwork	D	✓	✓	✓	✓	40	✓	
✓	5.	Insitu-concrete, large pre-cast concrete	E	✓	✓	✓	✓	30	✓	
✓	6.	Masonry	F	✓	✓	✓	✓	101	✓	
✓	7.	Structural Carcassing	G	✓	✓	✓	✓	68	✓	
✓	8.	Cladding and Covering	H	✓	✓	✓	✓	81	✓	
✓	9.	Waterproofing	J	✓	✓	✓	✓	20	✓	
✓	10.	Linings, Sheathing and Dry Partitioning	K	✓	✓	✓	✓	106	✓	
✓	11.	Windows, Doors & Stairs	L	✓	✓	✓	✓	128	✓	
✓	12.	Surface Finishes	M	✓	✓	✓	✓	289	✓	
✓	13.	Furniture and Equipment	N	✓	✓	✓	✓	106	✓	
	14.	Building fabric sundries	P	✓	✓	✓	✓	80	✓	
	15.	Pavings, Planting, Fencing and Site furniture.	Q	✓	✓	✓	✓	94	✓	
	16.	Disposal systems	R*	✓	✓	✓	✓	180	✓	
	–	Transport systems	X	–	–	–	–	–	–	Not applicable.
	17.	Services	R* - U W, Y	✓	✓	✓	✓	202	✓	
	–	SPARES						107		
		* Part						1652		

on each page, which can only be done by the typist. It does seem an advantage to have all the resulting slips (sometimes thousands) serially numbered, and, if there is any cross–referencing to be done, e.g. with spot items, this numbering is useful as the references can be inserted before typing. Some surveyors use a numbering machine to number the back of the slips for this purpose.

CHECKING

It will be found that special care is necessary in checking the typing. It is not so easy to spot mistakes in the cut and shuffle dimensions as in a draft bill written in the traditional form (though that is difficult enough!). Further information on checking the bill is given in Chapter 23.

INSTRUCTIONS TO THE TYPIST

Instructions to the typist must be precise, particularly as to the system adopted for marking headings, the extent of their continuation and the grouping of items. It may be found helpful to number all sheets to be typed (including preambles, headings, etc.) serially by numbering machine before handing to the typist. This gives a certain check if a sheet goes astray and helps the typists if they want to divide the work and afterwards reassemble.

PROGRESS SCHEDULE

A progress schedule will be found useful, so that at any time the stage reached by each section of the taking–off can be seen. A specimen schedule is shown on the previous page. The sections on the left hand side are taking–off sections, those on the right hand side are bill sections.

FILING

Cardboard boxes are probably best for keeping the slips and uncut sheets together during the progress of the contract, making them easily portable. If A4 paper is used for the uncut sheets, a standard box will probably be found suitable. If not, boxes can be made to special design economically if 50 or more are ordered at a time.

4 Applied mensuration

MATHEMATICAL KNOWLEDGE

It is assumed that the reader is acquainted with mensuration, knowledge of which, as of building construction, is an essential preliminary to a study of quantity surveying. In actual fact it is comparatively rarely that knowledge is required more abstruse than of the properties of the rectangle, triangle and circle. When a case does arise needing some less known formula, (see Appendix 3) or it can always be looked up in an appropriate mathematical reference book. The properties of the rectangle, triangle and circle must, however, be thoroughly understood, and if a student is not acquainted with them and with elementary trigonometry, these should be studied before going any further. This chapter shows some examples of how the theoretical knowledge of mensuration is applied to building work. Wherever possible lengths should be found by calculation from figured dimensions on the drawings, rather than by scaling. Where scaling has to be used a check should be made to ensure that other figured dimensions are accurate, as some reproduction methods affect the scale.

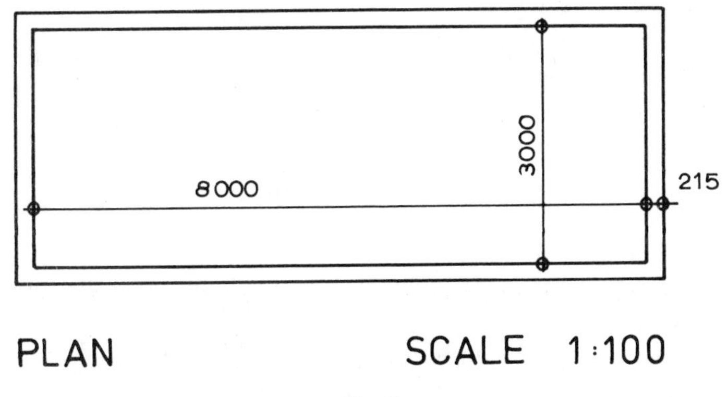

Fig. 1

PERIMETER OF BUILDINGS

The most common calculation required is the external and internal perimeter of a building and the length of the external walls which is required for the measurement of the foundations and brickwork. Figured dimensions may be given internally or externally, and either can be worked from. Figure 1 shows a plain rectangular building with one brick walls 215 mm thick:

The length or perimeter of the inside face of the walls which may be required for the measurement of finishings will be

```
2/8000 = 16000
2/3000 =  6000
         22000

OR        8000
          3000
       2/11000 = 22000
```

The length or perimeter of the inside face of the walls which may be required for the measurement of rendering can be calculated as follows:

```
        3000          8000
2/215    430           430
        3430          8430

2/3430 =  6860   OR   8430
2/8430 = 16860        3430
         23720     2/11860 = 23720
```

This may also be calculated from the internal perimeter as follows:

```
            22000
4/2/215 =    1720
            23720
```

It will be seen that twice the thickness of the wall has been added for each corner.

To arrive at the actual length of the wall, twice the external dimension in one direction may be added to twice the internal dimension in the other direction as follows:

$$\begin{aligned} 2/8\,430 &= 16\,860 \\ 2/3\,000 &= \underline{6\,000} \\ &\ \underline{\underline{22\,860}} \end{aligned}$$

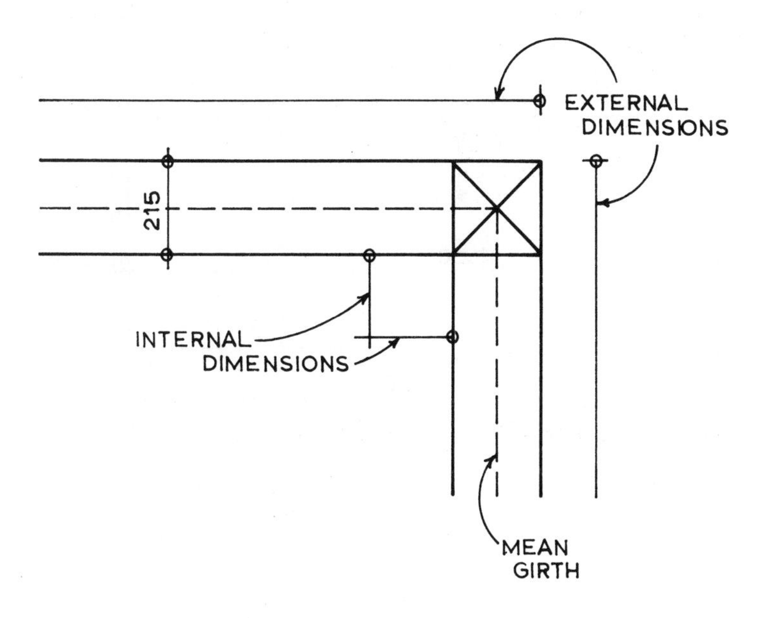

Fig. 2

A more straightforward way of calculating the above length is to find the centre line measurement or mean girth as follows:

$$\begin{aligned} &\ 22\,000 \\ 4/2/\tfrac{1}{2}/215 &= \underline{860} \\ &\ \underline{\underline{22\,860}} \end{aligned}$$

It will be seen that twice of half the wall thickness is added to the internal perimeter for each external corner.

The whole process can of course be reversed, if the external perimeter is taken instead of the inside, by deducting instead of adding, the mean length of the wall being obtained thus:

$$
\begin{array}{lrr}
 & & 23\,720 \\
\text{Less } 4/2/\tfrac{1}{2}/215 & = & \underline{860} \\
 & & \underline{\underline{22\,860}}
\end{array}
$$

If the shape of the building is slightly more complicated as shown in the diagram below the same principle may be applied to the calculations.

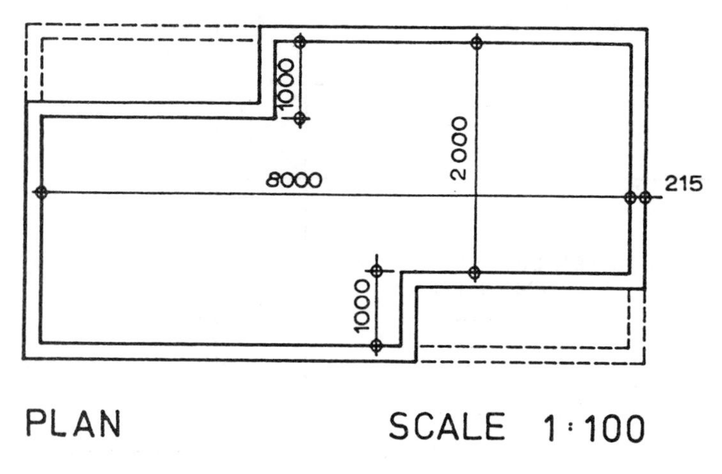

Fig. 3

For example the centre line or mean girth measurement is:

$$
\begin{array}{lrr}
2/3\,000 & = & 6\,000 \\
2/8\,000 & = & \underline{16\,000} \\
\text{Inside face} & & 22\,000 \\
4/2/\tfrac{1}{2}/215 & = & \underline{860} \\
\text{Mean girth} & & \underline{\underline{22\,860}}
\end{array}
$$

A check on the calculation could be made using the internal and external dimensions as follows:

$$
\begin{array}{rcl}
8\,000 & & \\
\underline{3\,430} & & \\
2/11\,430 & = & 22\,860
\end{array}
$$

Where the wall breaks back it will be seen from the following enlarged plan that the internal and external angles balance each other.

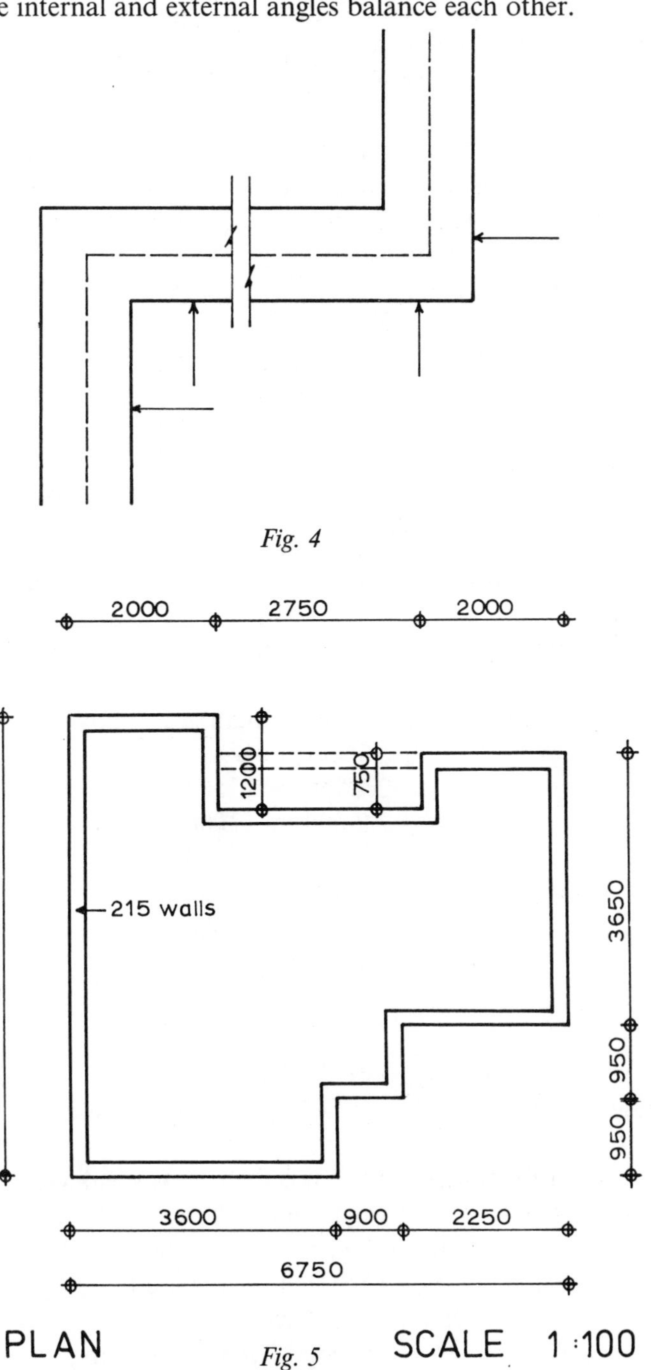

Fig. 4

Fig. 5

As before, the inside face having been measured, the external angle needs once the thickness of the wall added to give the length on the centre line, whereas in the case of the internal angle once the thickness of the wall must be deducted. In fact the perimeter of the building is the same as if the corners were as dotted on Figure 3. In short, to arrive at the mean girth once the thickness of the wall must be added for every external angle in excess of the number of internal angles.

This collection of the perimeter of walls being of great importance, one further and more complicated example is given in Figure 5, the calculations this time being made from the external figured dimensions instead of from the internal.

When smaller dimensions are given the first move should be to check that their total equals the overall dimensions given e.g.:

2000	3600		950
2750	900		950
2000	2250	1200	3650
6750	6750	− 750	450
			6000

Then the calculation of the mean length is as follows:

2/6750	=	13500
2/6000	=	12000
		25500
2/750	=	1500
		27000
Less $4/2/\frac{1}{2}/215$	=	860
		26140

If one imagines making a circuit of the walls, the breaks in the previous example are, so to speak, on the way round, but in this case it will be seen that at the top of the plan there is a re–entrant portion, the break of 1200 being only partly on the way round, as the next break of 750 goes back again and goes out of one's way. If the wall were as shown by the broken lines, then no addition would be necessary for the re–entrant portion, and the mean girth would be 24640, but as it is the two lengths of the 750 re–entrant breaks must be added. A check of the above calculation can be made by collecting up each length of wall, starting, say at the top left–hand corner and working round clockwise. In order to calculate the passings at angles correctly the usual method is to measure all horizontal lengths to extreme length, i.e. across the return wall, and all vertical walls between the horizontal ones only. By horizontal and vertical walls are here meant walls depicted by lines drawn horizontally or vertically on the paper. The check would be as follows:

2000	
985	(1200 – 215)
3180	(2750 + 2/215)
535	(750 – 215)
2000	
3220	(3650 – 2/215)
2465	(2250 + 215)
735	(950 – 215)
1115	(900 +215)
735	(950 – 215)
3600	
5570	(6000 – 2/215)
26140	

This method of collection would in any case be used if the building were of irregular shape, but where its angles are all right angles the length can usually be calculated from overall dimensions, though such piecemeal collection as this is useful as a check. In difficult cases it may be found convenient to mark by pencil lines on the plan the limits of each dimension.

The calculation of the mean girth is a most important one, as, having once been made, it is often used not only for several items in foundation measurement but also for brickwork and facings, with possibly copings, string courses, etc.

IRREGULAR AREAS

The triangle is constantly in use. Any irregular–shaped area to be measured is usually best divided up into triangles, each triangle being measured individually and added to give the area of the whole. If one of the sides, as for instance in the case of yard paving, is irregular or curved, the area can still be divided into triangles by the use of a 'compensating'

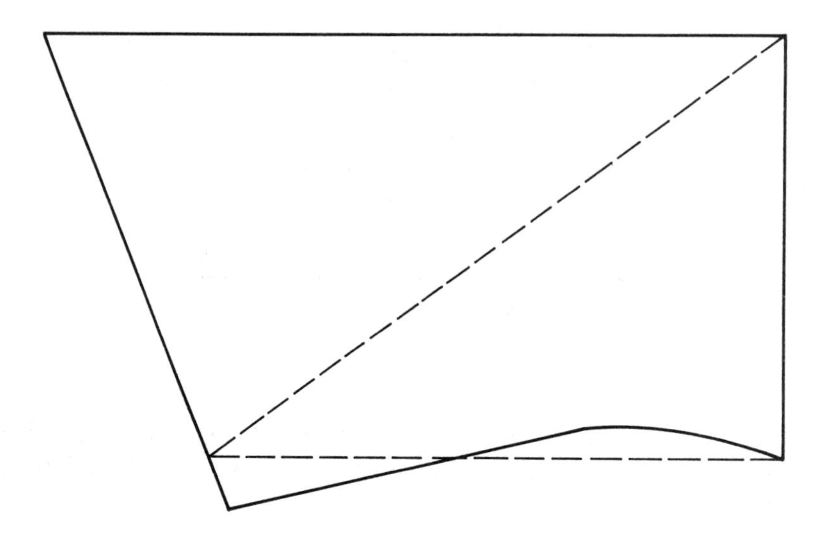

Fig.6

or 'give and take' line, i.e. a line is drawn along the irregular or curved boundary in such a position that so far as can be judged the area of paving excluded by this line is equal to the area included beyond the boundary.

In Figure 6 the area of paving to be measured is enclosed by firm lines, the method of forming two triangles (the sum of the areas of which equals the whole area) being shown by broken lines.

For a more accurate calculation of the irregular area, particularly if evenly spaced offsets are available dividing the area into an even number of strips, Simpson's rule may be applied. The intermediate offsets should be numbered as it is necessary to distinguish the odd numbers from the even. The formula is:

$$\frac{L}{3}(A + Z + 2E + 4Y)$$

Where L = Distance between offsets
A = Length of first offset
Z = Length of last offset
E = Sum of length of even offsets
Y = Sum of length of odd offsets

Where two sides of a four-sided figure are parallel to form a trapezoid it will not be necessary to divide into triangles, as the area is the length of the perpendicular between the parallel sides multiplied by the mean

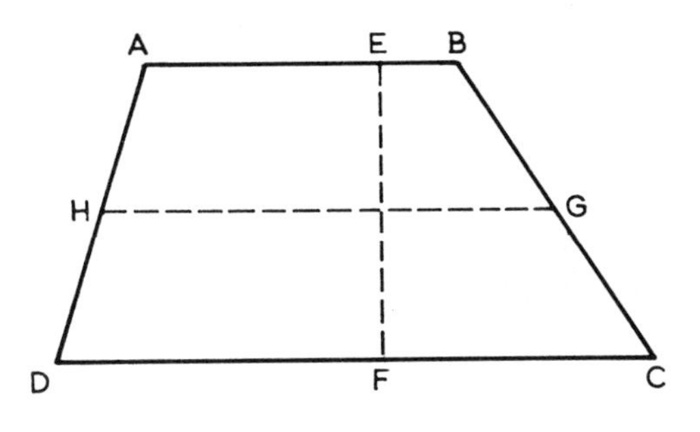

Fig. 7

length between the diverging sides. Thus, in this figure the area is EF × GH, GH of course being drawn half-way between AB and CD and being equal to

$$\frac{AB + CD}{2},$$

i.e. the average of AB and CD.

Another irregular figure which often puzzles the beginner is the additional area to be measured where two roads meet with the corners rounded off to a quadrant or bellmouth. This is most easily calculated as a square on the radius with a quarter circle deducted. For example:

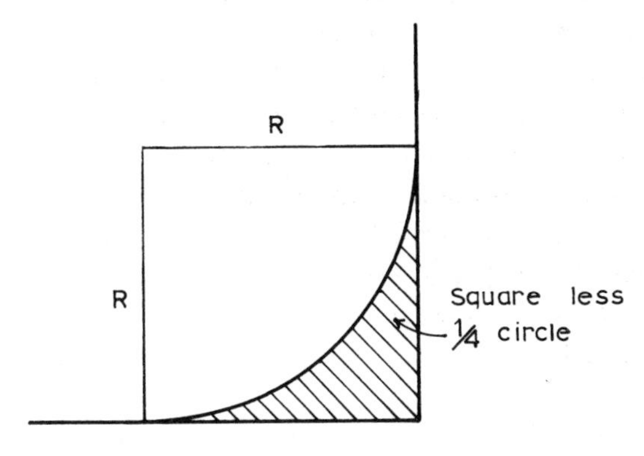

Fig. 8

$$\begin{aligned} \text{Additional area} &= R^2 - \tfrac{1}{4}\pi R^2 \\ &= R^2 - \frac{11}{14} R^2 \\ &= \frac{3}{14} R^2 \end{aligned}$$

ROOF SLOPES

The length of a roof slope too can be calculated from the triangle either for measurement or as a check where the length is scaled on a small drawing. The roof slope of a pitched roof is the hypotenuse of a triangle, the base of which is half the roof span and the height the height of the roof.

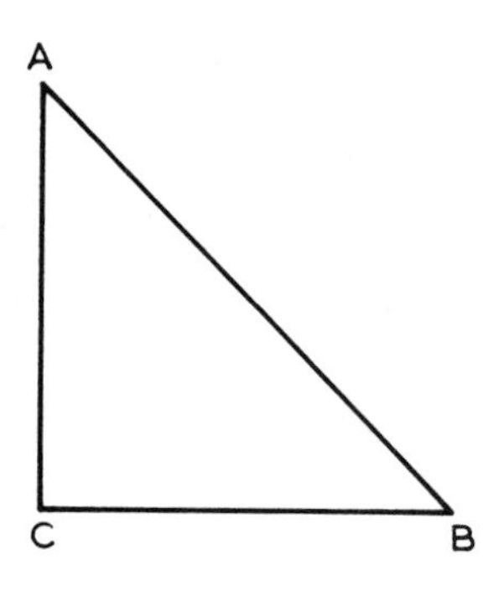

Fig. 9

In the triangle ABC, AB = BC × secant ∠ ABC, and if the angle of the pitch of roof is known the slope can thus be calculated. For example in the following:

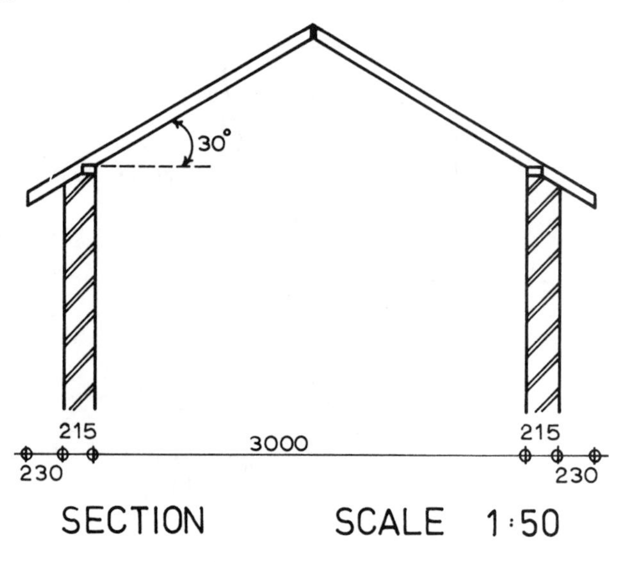

Fig. 10

half the roof span must of course be calculated to the extreme projection of the eaves or the edge of the tiling if coverings are being measured. The span inside walls being 3000, half the span for this purpose will be 1 500 + 215 + 230 = 1 945. The angle of pitch being 30°, the length of the slope will be

$$\begin{aligned} & 1\,945 \text{ secant } 30^\circ \\ = & \ 1\,945 \times 1.1\,547 \\ = & \ 2\,246 \text{ mm.} \end{aligned}$$

and this can be checked by scaling.

It should be noted that this calculation gives the length along the top of the rafter to the centre line of the ridge. This may have to be adjusted for the length of the covering which may project into a gutter or for the length of the rafter which should be the extreme length.

HIPS AND VALLEYS

The length of a hip or valley in a pitched roof must be calculated from a triangle, there usually being no true section through it from which it can be scaled. When the pitch of the hipped end is the same as the main roof then the length of the hip may be found from half the span and the length of the roof slope as calculated above.

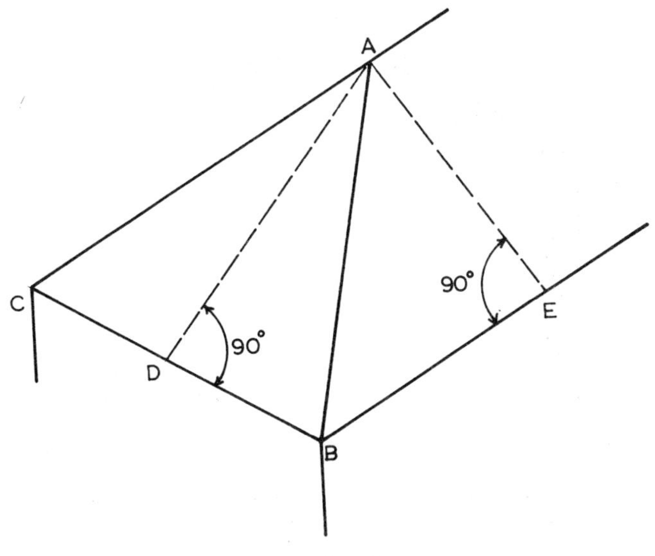

Fig. 11

For example in Figure 11:

BE and BD $=$ ½ span
AD and AE $=$ Length of slope
$\therefore$ $AB^2 = AD^2 + BD^2$

BROKEN-UP ROOFS

It should be noted that however a roof is broken up by hips and valleys, so long as the angle of pitch is constant the area will be the same, apart from any adjustments of projections of eaves, verge, etc. The area of tiling on a roof hipped at both ends may therefore be measured in the same way as if it were gabled, i.e. the length multiplied by twice the slope, the only difference being that if it were gabled the dimension of the length would probably be smaller, the projection of verges being less than that of eaves.

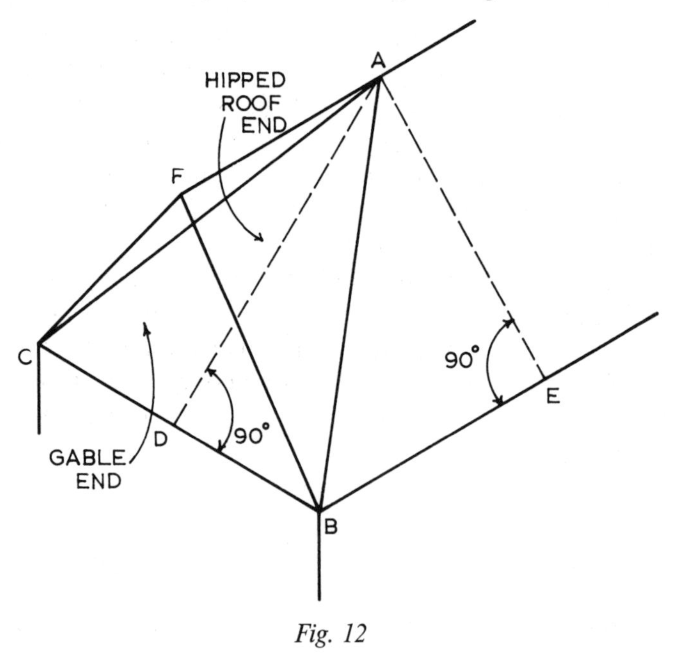

Fig. 12

For example in Figure 12:

The area of triangle ABD = the area of triangle ABF
BE = BD = AF = ½ span
AB is common

$\angle$ ADB = 90° and $\angle$ AFB = 90°
$\therefore$ with two sides and one angle equal, the triangles
ABF and ABD are equal in area

Just as the length of slope is the length on plan multiplied by the secant of the angle of pitch, so the area of a roof of constant pitch is the area on plan multiplied by the secant of the angle of pitch, and this formula can be a useful check on the measurements when worked out.

MEASUREMENT OF ARCHES

Where arches are to be measured or the wall below deducted, the measurement is fairly straightforward for semi-circular arches. In the case of segmental arches the deduction above the springing line and the girth of the arch are not usually calculated precisely, as they can be judged sufficiently accurately, the former by a triangle with compensating lines or by taking an average height, and the latter by stepping the girth round with dividers. In the case of expensive work, one should be as accurate as possible, the measurements preferably being worked out by calculation. A rough method of measuring the area of a segment is to take $^{11}/_{16}$ times the area of the rectangle formed by the chord and the height of segment. This is obviously not mathematically correct, and the margin of error will vary with the radius and length of chord, but when dealing with small areas this method will often be found sufficiently accurate. Another way is to take first the inscribed isosceles triangle based on the chord, leaving two much smaller segments each of which may be scaled and set down as base × $^2/_3$ height. The error is then very small only. In dealing with large areas of expensive materials a more accurate method would be necessary using the formula

$$\frac{H^3}{2C} + (^2/_3 C \times H)$$

where C is the length of chord and H the height (see Chapter 2).

EXCAVATION TO BANKS

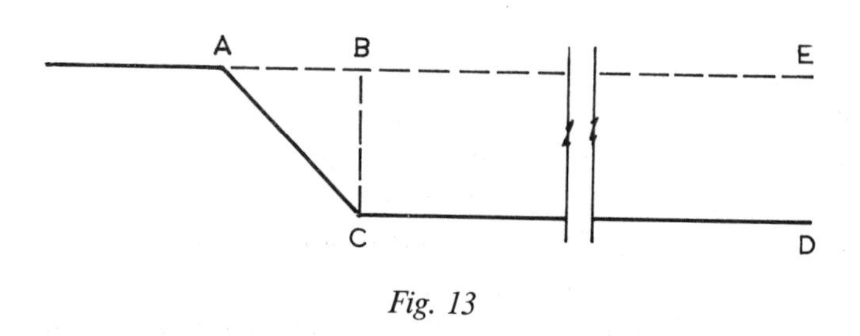

Fig. 13

The volume of excavation necessary on a level site to leave a regular sloping bank is naturally the sectional area of the part displaced (triangle ABC above) multiplied by the length, the volume of the remaining excavation being the sectional area of the rectangle BCDE also multiplied by the length. It may be, however, that the natural ground level is falling in the length of the bank as shown in the following diagram:

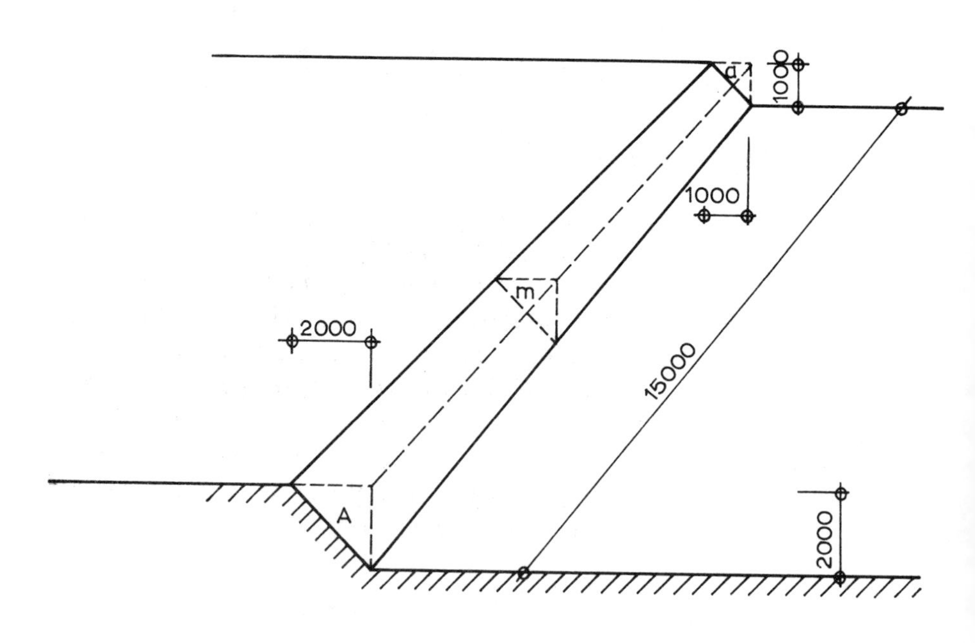

Fig. 14

The volume of earth to be displaced should then theoretically be calculated by the prismoidal formula. As the final volume is required in cubic metres the calculation is carried out in metres.

$$V = \frac{L(A + a + 4m)}{6}$$

where V is the volume,
L is the length,
A is the sectional area at one end,
a is the sectional area at the other end,
m is the sectional area at the centre.

Note that the area m is not the average of areas A and a, but should be calculated from the average dimensions.

If AB and BC in Figure 13 are both 2.00 m at the higher end and 1.00 m at the lower end they would, assuming a regular slope, be 1.50 m at the centre. The volume of the prismoid in Figure 14 in cubic metres would therefore be:

$$\frac{15\left(\frac{2 \times 2}{2} + \frac{1 \times 1}{2} + 4 \times \frac{1.5 \times 1.5}{2}\right)}{6}$$

$$= \frac{15(2 + 0.50 + 4.50)}{6} = \frac{105}{6}$$

$$= 17.50 \text{ m}^3$$

In practice, however, it will often be found that so precise a calculation is not made in such cases. The surface of ground, whether level or sloping, is not like a billiard cloth, and the natural irregularities prevent the calculation from being exact. Moreover, in dealing with normal building sites the excavation for banks is a comparatively small proportion of the whole excavation involved (unlike, say, the case of a railway cutting), and any departure from strict mathematical accuracy due to the use of less precise methods would only involve a comparatively small error. In the example given above the volume might in practice be taken as the length multiplied by the sectional area at the centre, i.e.:

$$15 \times \frac{1.50 \times 1.50}{2} = 16.88 \text{ m}^3$$

Although an error of 0.62 m^3 may be thought high, it must be remembered that a case is being assumed where the excavation to the bank is itself only a small proportion of the whole.

If the natural ground level falls to such an extent that it reaches the reduced level, and the bank shown in Figure 14 therefore dies out to nothing, thus:

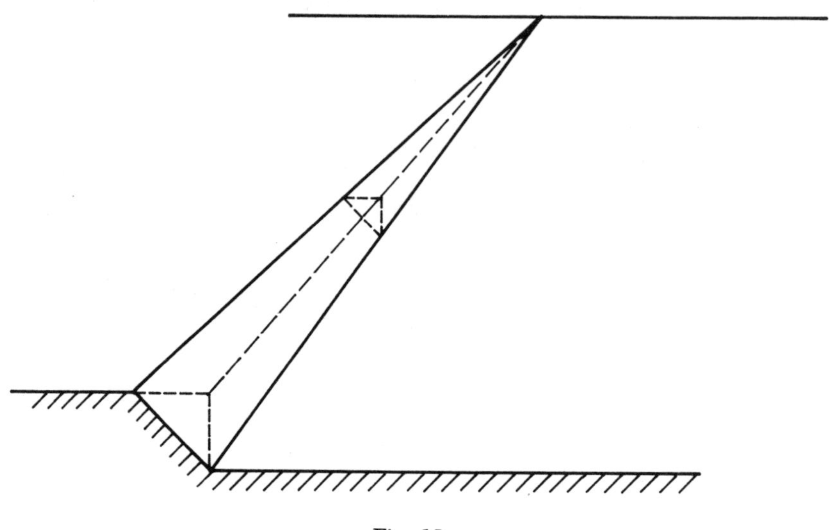

Fig. 15

the formula will be found to simplify itself.

a becomes zero,

m becomes by simple geometry $\frac{A}{4}$, because the triangles at A and m are right–angled triangles, that at m having two sides enclosing the right angle each half the length of the corresponding sides of the triangle at A.

$$V = L \frac{(A + 0 + 4 \times \frac{A}{4})}{6} = \frac{L \times A}{3}$$

which is the formula for volume of a pyramid.

If the volume of earth to be excavated forms an even number of prismoids of equal length then Simpson's rule may be applied taking the area at the offsets rather than the length as in the case of the irregular area in Figure 14.

EXCAVATION TO SLOPING SITES

The theoretical principle for measuring the volume of excavation in cutting for a sloping bank may be extended to a sloping site.

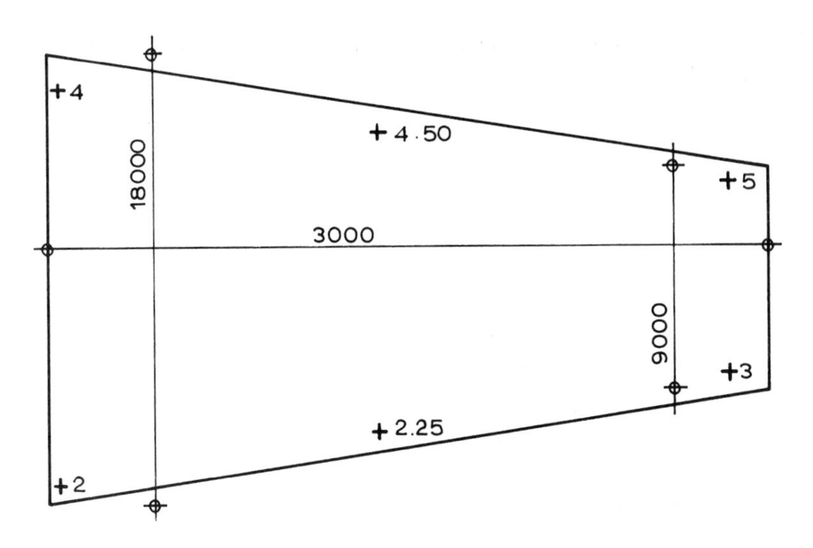

Fig. 16

If the figures given above are natural levels and it is required to excavate to a general level of +1 (the end boundaries of the area being parallel), the volume of excavation may be calculated by the prismoidal formula as already given of

$$V = \frac{L\,(A + a + 4m)}{6}$$

When $L = 30$

$$A = 18 \times \frac{3 + 1}{2} = 36$$

$$a = 9 \times \frac{4 + 2}{2} = 27$$

$$m = 13.50 \times \frac{3\tfrac{1}{2} + 1\tfrac{1}{4}}{2} = 13.50 \times \frac{19}{4} \times \tfrac{1}{2} = \frac{256.50}{8}$$

$$\text{then } V = \frac{30\left(36 + 27 + \frac{256.50 \times 4}{8}\right)}{6} = \frac{30(36 + 27 + 128.25)}{6}$$

$$= 5 \times 191.25 = 956.25 \text{ m}^3$$

If only a simple average of the four corners were taken the result would be:

		4
		5
		3
		2
	÷ 4)	14
Average ground level		3.50
Reduced level		1.00
Average depth		2.50

$30 \times 13.50 \times 2.50 = 1012.5\,m^3$

This shows an error which is greater because the intermediate level on the lower boundary is not an average of the two end levels.
An average of the eight levels (giving the centre levels double value) would be:

		4
2/4.50 =		9
		5
		3
2/2.25 =		4.50
		2
	8)	27.50
		3.44
		1.00
		2.44

or a volume of $30 \times 13.50 \times 2.44 = 988.2\ m^3$, from which it will be seen that there is a definite error. This method would be satisfactory, however, if the area of the ground were a rectangle.

The error would vary with the regularity of the slope, and where the slope is fairly regular it may often be sufficiently accurate to take an average depth over the whole area. For an ordinary building site, calculation would probably be made in this way in practice.

If it is required to apply the formula where end boundaries are not parallel it will be necessary to draw parallel compensating lines along these two boundaries. It is of course assumed that slopes between the levels given are regular. If a line of intermediate levels were given between the upper and lower line in the diagram above, it would be necessary to treat the two portions separately.

GRIDS OF LEVELS

On larger sites it is customary to take a grid of levels over the whole area, or at least over the area of the proposed construction, at regular intervals forming squares plus additional levels at any significant points such as existing manhole covers or along embankments. When calculating the average level of an area covered by a grid it is necessary to find the average level of each square of the grid by totalling the levels at the four corners and dividing by four. Each of the average levels thus found are added together and this total is divided by the number of squares and the result gives the average level of the area. To calculate the amount of excavation the total area is multiplied by the difference between the average level of the area and the formation level required. If the excavation of top soil has been measured as a separate item then the depth of this would have to be deducted from the average level.

A quicker method is to total the levels in each of the categories indicated below and multiply each total by the appropriate 'weighting' shown.

(a) Levels at the external corners on the boundary of the area	Multiply by 1
(b) Levels at the boundaries of the area (other than those in (a) or (c))	Multiply by 2
(c) Levels at any internal corners on the boundary of the area	Multiply by 3
(d) Intermediate levels within the area	Multiply by 4

The results thus obtained are totalled and divided by a number equal to four times the number of squares (or the total number of weightings). The result is the average level of the area and will produce the same result as the first method.

Sometimes a site may be partly excavated and partly filled and it will be necessary to plot a 'cut and fill' contour using interpolation as shown below to ascertain the location. Contours may also have to be plotted to represent the division between depth bands of excavation or fill as required by the SMM. Grid squares which are cut by the contour line to form triangles or trapezia should be dealt with separately, the average level of each is found by adding the levels at the corners and dividing by the number of corners. The area of each irregular figure is multiplied by the difference between its average level and the formation level to give the volume of excavation or fill to add to the main quantity.

When measuring excavation work one should always look for sudden changes in levels which may indicate items such as embankments or craters. These areas should be dealt with separately and not averaged within the main area as above.

INTERPOLATION OF LEVELS

When measuring excavation work it is sometimes necessary to ascertain the ground level at a point between two levels given on the drawing or to locate a point at which a certain level occurs. This can be achieved by interpolation as shown in the following example:

Given two levels at points A and B 50 m apart and requiring to find an intermediate level at a point C 20 m from A:

Level at A = 3.60

Level at B = $\underline{2.90}$

Difference 0.70

Therefore the ground falls 0.70 over 50 m

To find the fall y over 20 m

$$\frac{0.70}{50} = \frac{y}{20}$$

$$20 \times 0.70 = 50y$$

$$y = 0.28$$

The level at C = 3.60 – 0.28 = $\underline{\underline{3.32}}$

If the intermediate level is known, such as a formation level, then the distance from A could be found in a similar manner, the unknown being the distance rather than the level.

OTHER CALCULATIONS

Various other lengths, areas and volumes may have to be calculated and reference to mathematical formulae, a selection of which is given in Appendix 3, will indicate how to make the appropriate calculation.

5 General rules for taking–off

In this chapter are given general rules of measurement and other information that is considered necessary for the study of the examples which follow. The general organisation and procedure in taking–off dimensions of a complete building are dealt with in Chapter 19, but lack of knowledge of these need not keep the reader from proceeding with the examples.

SECTIONS OF TAKING–OFF

The taking–off of dimensions is usually divided into sections under three main subdivisions:

(a) Carcase (b) Finishings (c) External works

The sections found in a normal building, in the order in which they will be dealt with in this book, are as follows:

(a) Carcase

(1) Foundations
(2) External walls
(3) Internal walls
(4) Fires and vents
(5) Floors
(6) Roofs

(b) Finishings

(7) Internal finishings
(8) Windows
(9) Doors, including openings without joinery
(10) Fixtures and sundries
(11) Stairs and lifts
(12) Plumbing and engineering services

(c) External Works
(13) Drainage
(14) Roads, paths, site layout, outbuildings etc

A further sub–division for alterations and repair works sometimes known as 'spot items' may be required.

The terms 'carcase' and 'finishings' cannot be taken in an absolutely literal sense, as there must necessarily be a certain amount of overlapping. Some might consider external facings and roof coverings as 'finishings', but, being part of the structure, they are usually included in the carcase. Floor finishings may be measured with the floor construction, but some surveyors prefer to take them with the internal finishings. The order of sections too may follow the particular taste of the surveyor, but it is best, having decided on a definite order, to follow it always. It must be understood that the special requirements of any building may require additional or subdivided sections, and the list given above must therefore be regarded as elastic.

TAKING–OFF BY WORK SECTIONS

Some surveyors make a practice of taking–off by work sections instead of by parts of the building as described above. As the final bill is usually arranged in work sections, this system can eliminate the intermediate slip sorting or abstract, which, when measuring by sections of the building, is necessary to collect and classify the items into work sections. Such a system may present difficulties in some cases, and would seem to increase the risk of overlapping or of forgetting something. When measuring by sections of the building the taker–off mentally erects the building step by step and is less likely to miss items. Sometimes a combination of the two systems may be used to advantage.

OVER–ALL MEASUREMENTS

It is usual in measuring to ignore in the first instance openings, recesses and other features which can be dealt with by adjustment later. Brickwork, for instance, is measured as if there were no openings at all, and deductions are made when the windows, doors or other openings are dealt with in their proper section. Plastering and similar finishings are measured in the same way. It simplifies the work to measure in this way, as when, for instance, windows are being considered the sizes will be to hand, and openings will be measured to correspond. Moreover, it may happen that, say, internal plastering and windows are being dealt with by different takers–off, when it is obvious that the one who measures the windows is better able to make the deductions and adjustments. Measuring over all with adjustment later is preferable to piecemeal measuring. This principle will further be found of value if a window should be forgotten. The extra cost of window over the wall and finishings would alone be involved by the error, a much less serious matter than if nothing at all had been measured over the area of window.

USE OF SCALES

A warning should be given of the possibility of using the wrong scale in measuring. If possible, each side of the scale used should not have more than one variety of marking on each edge, but this is not always practicable. The scale most easily available for general use is a standard metric scale having 1:5, 1:50, 1:10, 1:100, 1:20, 1:200, 1:1250, and 1:2500 markings. Some surveyors prefer to have a separate scale for each variety, or one marked on one face only, but when working on two different drawings, e.g. 1:100 and 1:20, at the same time – as is quite common, it is a great convenience to have both markings on the same scale. In any case special care is necessary when measuring from different drawings to see that readings are taken from the correct markings of the scale. It may seem unnecessary to emphasise this, but mistakes on this account are not unknown, sometimes even by the draughtsman.

ABBREVIATIONS

As mentioned above a list of abbreviations in common use is set out in Appendix 1, but this list is in practice considerably extended by the shortening of words generally and by other abbreviations understood in the particular office, or in the context of the whole description. In the case of cut and shuffle taking–off such abbreviations are restricted to slave slip descriptions(see Chapter 3).

A special note might perhaps be made here of the abbreviation 'a.b.' for 'as before'. Where this is used and might refer to more than one item it always refers to the last such item. For instance, a description '40 mm door a.b.' would refer to the last type of 40 mm door measured, if there have been several different varieties. If however, there is any doubt it is best to add to the brief description sufficient for it to be identified or to say '40 mm door a.b. col. 146', the reference to the column number being a definite guide. The use of 'ditto' or 'do' when used in billing is dealt with in Chapter 21.

DESCRIPTIONS

The framing of descriptions so that they are both clear and concise is an art not easily acquired, but one which is of the utmost value. The builder's estimator, always working at high pressure, will waste much time if faced with long–winded and rambling descriptions, or having to decide what the surveyor intended to convey by a confused sentence. The surveyor, therefore, must always aim at clear expression being careful in the choice of words, and using the various technical terms in their proper sense. When using standard libraries the descriptions are in the main laid down for the taker–off. However, takers–off have to be very careful to avoid the pitfall of fitting an item to a standard description rather than ensuring that the description fits the item; if it does not then a special or 'rogue' must be compiled.

The requirements of the SMM should be followed carefully when framing descriptions, but it must be remembered that additional information should be given where necessary to convey the exact nature of the work to the estimator (SMM GR 1.1). Whilst it may be convenient to follow the order of the tabulated rules of the SMM one is not prevented from using traditional prose in the framing of descriptions (SMM Preface second paragraph). Certain items such as waste of materials, square cutting, fitting or fixing materials or goods in position, plant and other items listed in Clause GR 4.6 of the SMM are generally deemed to be included in descriptions. Where the SMM calls for a dimensioned description to be given, apart from the description of the item, all dimensions should be given to enable the shape of the item to be identified (SMM GR 4.7).

The wording of descriptions is dealt with more fully in Chapter 21 (Bill Preparation), but descriptions well drafted on the dimensions in the first instance will simplify considerably the work of the editor. The taker–off must also be careful to see that the same wording is used when referring to the same thing in different parts of the dimensions, as inconsistency in the descriptions may indicate that there must be some distinction intended by the different phraseology. For instance, if the taker–off, describing plaster writes '2 Coat plaster on block walls' and then after several such items writes later '2 coat plaster on partitions' they may end up as different items in the bill although this was not the intention of the taker–off. Therefore, when the same item appears in different places it should be written in exactly the same form, or after the first time abbreviated with the letters a.b. to indicate that exactly the same as before is intended. It is important to confine the description to what is actually to appear in the bill, and not to add to it particulars of location or other notes which are merely put for reference and not intended to go any further.

The descriptions written on the dimensions should be in the form that they are intended to appear in the bill, such parts which would normally be covered by a preamble (see Chapter 22) being omitted. Notes to assist in writing the preambles, particularly for the less common items, can be entered on the dimension sheet but should be kept well clear of dimensions or descriptions. In practice, certain items which are constantly met with may have the description curtailed, being described by reference to the bill for some other job; the taker–off's time in writing out a long description is thus saved.

In some cases there will be added to the description of a superficial or linear item a note of the number included in the item so that the estimator can judge the average size of each, e.g.:

3/	0.22 0.40		4 mm clear sheet glass to wood with beads in panes not exceeding 0.15 m^2 (In Nr. 3 panes).

Some of the requirements of the SMM as to descriptions may be covered by general clauses or preambles to each work section. For example, SMM H71.S.4. says that the lap in joints of flashings shall be given. A preamble clause saying that all flashings are to be lapped 100 mm at joints would be sufficient and probably save repetition.

DIMENSIONED DIAGRAMS

The sixth edition of the SMM called for more drawn information to be made available than in previous editions and SMM7 has reinforced this call. The implications of these requirements are brought out in SMM and in the MC. Most of this drawn information will be available from the drawings used for the taking–off and it will only mean ensuring that the information required by the SMM is included and that the requisite number of copies of the various drawings accompany the tender documents. However, in certain cases the SMM suggests dimensioned diagrams or sketches to elaborate a description. These diagrams or sketches can either be produced separately for inclusion in the text of the bill of quantities or drawn by the taker–off with the dimensions to be processed as a separate operation. How these sketches appear in the bill of quantities is a matter for the individual surveyor to decide; see also Chapter 19.

'EXTRA OVER'

Some items are measured as 'extra over' others, that is they are not to be priced at the full value of all their labour and materials, as these have to a certain extent already been measured. For example fittings such as bends and junctions to drain pipes or angles and ends to gutters are measured as 'extra over'. This means that the pipe or gutter is measured its full length over the fittings and the estimator when pricing the item assesses the extra cost for the fittings substituting the original item. On small pipes, although fittings are measured as 'extra over' the saving of pipe is so minimal that it will probably be ignored by the estimator. Such items are simply described as 'extra for' where it is obvious what they are extra over, e.g., 'Extra for bend' following immediately after an item of pipe obviously means 'extra over the cost of the pipe in question for a bend'. In the bill these items may be 'written short' on the main item so that they are identified together. The measurement of an item as 'extra over' something already measured as 'extra over' should be avoided.

FORM OF CONTRACT

The forms of contract agreed by the Joint Contracts Tribunals provide that measurements shall be made in accordance with the SMM (see Chapter 1) and it is therefore of the utmost importance, where these forms of contract are used, that the standard method should be followed. It should, however, be understood that unless referred to in the contract the SMM has no legal sanction and need not be adopted. However, this document has now been so long established that it could be produced as evidence of custom in the profession, and it is therefore advisable to study it and to follow its recommendations in all cases, or make quite clear where there is a divergence.

MEASUREMENTS FOR ANALYSIS OF PRICE

A distinction must sometimes be made between the measurements necessary for the preparation of a bill of quantities and those necessary for the analysis of a price for a particular item. This distinction is often not realised, especially by students. It has been found by experience that in certain cases a price can be built up for a unit of measurement without pressing the analysis of the item to its extreme limit, with all the additional work involved. For example brickwork is measured including the mortar and the estimator who is used to pricing this item will have an accurate idea of the rate for the whole item and if not, the item can be analysed further to arrive at its value.

MEASUREMENT OF WASTE

It may be taken as a general rule (though like all rules it has exceptions) that measurements of work are made to ascertain the net quantities as fixed or erected in the finished building. Wastage of material generally is allowed for by the builder in the prices, though sometimes a measurement is made as a guide to the amount of waste, and in a few cases the gross quantity is measured. The exceptions to the general rule will be pointed out as they occur.

AVERAGING

Where items are similar and differ only in one dimension they may be taken together, the varying dimension being averaged. It is important to note that an average cannot strictly be taken in more than one dimension. This will be seen plainly in comparing the following.

	2.50 1.80	4.50	
	1.50 1.20	1.80	
		6.30	

which it might appear could be averaged as:

2/	2.00 1.50	6.00	

The difference in the above results shows the amount of error, which is not inconsiderable, and would be increased if a third dimension were added, as say in the case of the excavation for two manholes 1.50 M deep:

	2.50 1.80 1.50	6.75	
	1.50 1.20 1.50	2.70	
		9.45	

2/	2.00 1.50 1.50	9.00	

where the error is nearly ½ cubic metre for two manholes alone.

6 Substructures

PARTICULARS OF THE SITE

Before beginning the measurement of foundation work the drawings must be examined to ascertain whether the existing ground levels are shown sufficiently for calculating average depths of excavation. If the levels are not shown or are insufficient then it is necessary to take a grid of the levels over the site. Irrespective of taking levels the surveyor should always visit the site to ascertain the nature and location of existing buildings, details for preliminary items and for the measurement of excavation work. Amongst items to be noted for the latter are vegetation to be cleared, the existence of topsoil or turf to be preserved, pavings or existing structures in the ground to be broken up and, if trial holes have been dug, the nature of the ground and the ground water level. The visit to the site is often best left until the general taking–off is done, so that notes can be made during the taking off of what must be looked at on the site. If, however, the site is close at hand a quick visit before work in the office starts can often prove beneficial. Where, of course, the proposed work consists mainly of alterations, an early visit to the site will be necessary, and most of the taking–off may even have to be done there. The measurement of alterations or 'spot items' and the methods of dealing with this class of work are dealt with in Chapter 18.

BULKING

When measuring excavation, disposal and filling the dimensions are taken as they are required in the ground, trenches being measured vertically above the sides of the foundation. Soil increases in bulk when it is excavated but no account is taken for this in the bill of quantities, the estimator having to make the due allowance.

REMOVING TOPSOIL

Where new buildings are to be erected on natural ground it is required to measure separately a superficial item for the stripping of the vegetable or topsoil where it is to be preserved. This is measured over the area of the whole building including the projection of concrete foundations beyond external walls. A separate cubic item has to be taken for the disposal of the topsoil giving the location. Any further excavation for trenches, basement, etc., would then be measured from the underside of such topsoil excavation (formation level). If there were some existing concrete paths, tar paving, etc., over portions of the area to be stripped, an item must be taken for breaking out existing hard pavings as a superficial item stating the thickness and the material and measuring as a separate item the removal. Breaking out may be taken as extra over the excavation.

Where the site is covered by existing buildings, the pulling down would be dealt with as described in Chapter 18, no item of stripping topsoil being necessary. Demolition is usually taken to existing ground level and excavation together with any necessary breaking out measured as a cubic item which may be taken as extra over the excavation.

It often happens that part of the site is covered with good turf which is worth preserving, and in such cases an item should be taken for lifting it (SMMD20.1.4.1) and a separate item for relaying any to be reused. (SMM Q30.4)

EXCAVATION OVER SITE

Where a site is sloping it is often more economical to set the ground-floor level so that one end of the site must be dug into, i.e. the underside of part of the hardcore bed will be below the level of the ground after the topsoil is stripped. Where this is the case, a cubic measurement is made of the excavating to reduce levels necessary from the underside of the topsoil excavation already measured to the underside of the hardcore bed. The depth for this item must be averaged, and it will generally be found that this digging is only ncessary over part of the site, the level of the rest being made up with hardcore filling. In Figure 17 it will be seen that the ground level must be reduced to 44.00 (300 mm below top of floor slab). The contour of 44.00, known as the cut and fill line, is plotted on the plan as accurately as possible from the levels given; the area on the right hand side of this contour is measured for excavation, that on the left hand side for filling. If topsoil has been dug to a depth of 150 mm then the contour plotted for the cut and fill line would be 44.15. Additional contours may have to be plotted to classify the excavation according to the maximum depths required by the SMM. It may, however, be more sensible to find an average depth for the whole excavation for measurement purposes and classify the depth in the description as the maximum on site. If this method is used a statement should be made in the bill to this effect.

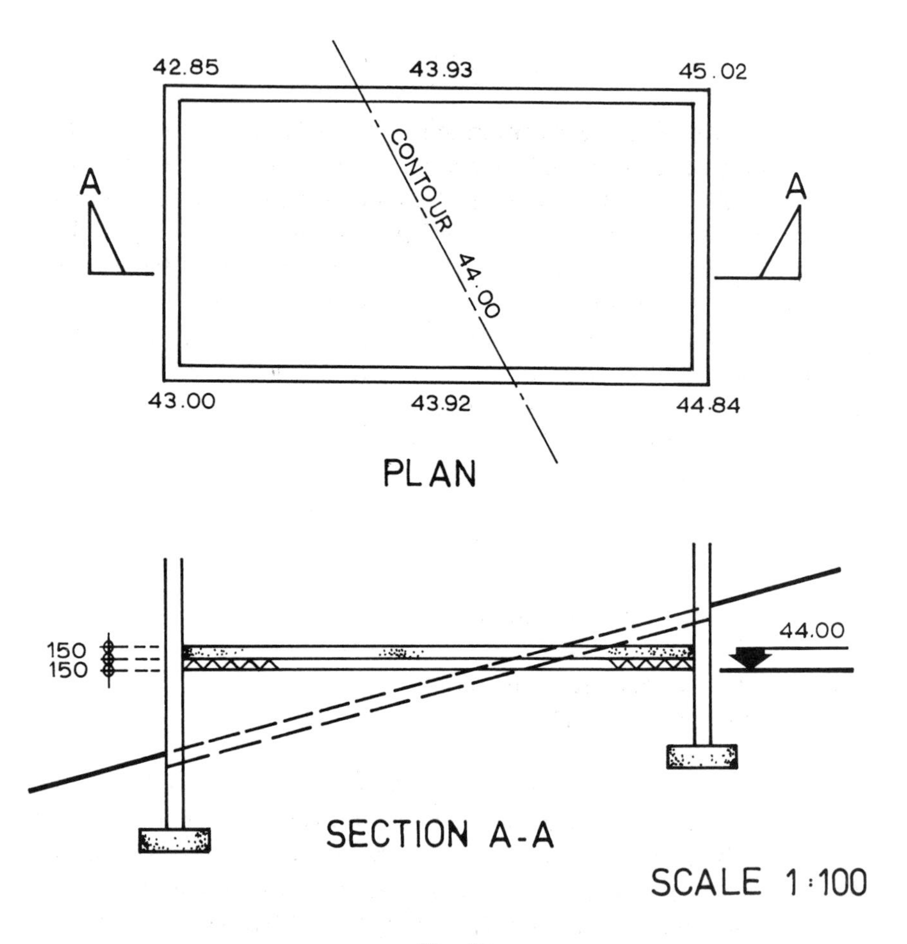

Fig. 17

The excavation to reduce levels must be measured before any trench digging, as it brings the surface to the 'reduced or formation level' from which the trench digging is measured, and, like the stripping of surface soil, it must be measured to the extreme projection of concrete foundations. A separate cubic item of disposal of excavated material must also be taken. It will often be necessary, where the floor level of the building is below the ground outside, to slope off the excavation away from the building, and possibly to have a space around the building excavated to below the floor level. In such circumstances the additional excavation would be measured with the external works section and a superficial item taken for trimming the side of the cutting.

DIGGING FOR PATHS

Stripping of topsoil and excavation to reduce levels may also be required for formation of paths, paved spaces, etc. External work of this nature is best measured all together after the building is dealt with, as it usually forms a separate section in the bill. When this is so, the paths, etc., abut the building (thus overlapping the projection of foundations), the whole excavation necessary for the erection of the building should be measured with the building, the extra width necessary for the paths only being measured with the paths. If this is done, and it is decided not to accept the estimate for paths, the full excavation necessary for the building will still be included in the estimate for the building.

LEVELS

Before foundations are measured, three sets of levels must be known:

(1) bottom of concrete foundation
(2) existing ground level
(3) floor level

(1) and (2) are necessary to measure trench excavation and (1) and (3) are necessary to calculate correct heights of brickwork or other walling. The natural ground level, as has been pointed out, will probably vary and have to be averaged either for the whole building or for sections of it, and if the floor level and the bottom of the foundation are constant, the measurement of trench excavation is fairly simple. However, both bottom of the foundation and the floor level may vary in different parts of the building, there being steps at each break in level, and sometimes the measurement of foundations becomes very complicated on this account. It will be found useful to mark on the plan the existing ground levels at the corners of the building, if necessary being interpolated from given levels and in the same way the levels of bottom of the foundation could be

marked on the foundation plan (if any). If stepped foundations are required it will prove helpful if the foundation plan is hatched with distinctive colours to represent the varying depths of the bottom of the foundations. If no foundation plan is supplied, the outlines of foundations can be superimposed on the plan of the lowest floor.

TRENCH EXCAVATION

The measurement of trench excavation and other foundation work will normally divide itself into two sections:

(a) external walls, and
(b) internal walls

the former being dealt with first. In the simplest type of building a calculation will be made of the mean length of the trench as described in Chapter 4. This mean length of the trench, will also be the mean length of the concrete foundations. The width of trench will be the width of the concrete foundation as shown on the sections or foundation plan. The depth of trench will, if the bottom of concrete foundation is at one level, be the difference between that level and the average level of ground, after making allowance for stripping of the topsoil or excavation to reduce levels already measured. Where the bottom of concrete is at different levels, theoretically the digging for each section of foundation between steps should be measured separately, the lengths when measured being collected and checked with the ascertained total length. It may be found, however, in practice that, the steppings to bottom of trench being small, and the ground normally falling in the same direction, an average depth can be decided on for larger sections of the building, if not for the whole. The depth of trench excavation will have to be stated in the description in accordance with the requirements of the SMM and, if the depths vary, it may be necessary to measure different parts separately to keep to the SMM classifications. Alternatively, as mentioned above, in excavation to reduce levels, the maximum depth may be given and a statement made in the bill giving the method used. The main measurement of excavation for the external walls having been set down, excavation for any projections on this foundation can be measured.

It has been assumed above that the external walls have trenches of a uniform width all round. If there are several different widths, as may be the case where the thickness of wall varies, each width would be dealt with separately, different lengths of the same width being collected together on waste, and the whole being carefully checked with the ascertained total length. Trench excavation is measured as a cubic item and a separate cubic item of disposal must be taken. Trenches not exceeding 300 mm wide are kept separately (SMM D20.2.5).

Internal walls must be collected up in groups according to the width of their foundation and the average depths. Allowance should be made in the length for overlap where an internal wall abuts against an external wall by deducting from the length of the internal wall the projection of concrete foundation to the external wall at this point. A similar allowance should be made where internal walls intersect. The necessity for this is best shown diagrammatically:

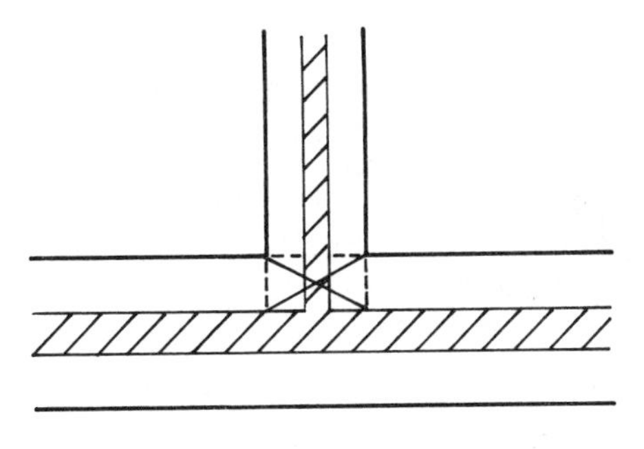

Fig. 18

It will be seen that if the foundation for the internal wall were measured the same length as the wall (i.e. as dotted) the area marked with a cross would be measured twice for the excavation and concrete. The amount involved being comparatively small, some surveyors may ignore the deduction, but to do so, if there are many intersections, is to take an unnecessary and definite full measurement without any really justifiable reason.

Students often find it difficult to decide whether the maximum depths for trench excavation, etc., should be calculated from the original ground level or from the level of the ground after topsoil has been stripped, where this is measured separately. It is usual to measure from the latter level as this is the commencing level of the actual excavation being measured. The SMM requires that the commencing level of excavation must be stated where this exceeds 250 mm from the existing ground level, the depth classification of excavation being given from the commencing level.

BASEMENT DIGGING

Basement digging will be measured from the underside of the stripping of the topsoil or from the reduced level down to the underside of the basement floor, the trenches below being measured separately as trenches from the basement level. Where only part of the area of a building has a basement, it will be found most convenient to measure first all the basement complete up to a certain level, say ground floor or general damp proof course and afterwards to measure the remaining foundations up to the same level.

EARTHWORK SUPPORT

Earthwork support must be measured as a superficial item to the face of the excavation to be upheld, whether it will actually be necessary or not, to cover the builder's responsibility to uphold the sides. It is for the builder to decide, from the information as to the nature of the soil, the extent and strength of support that will be required. Even if the decision is not to use any, there is still a risk to be priced; if the excavation falls in, it will have to be re–excavated at the contractor's own expense. The surveyor measures the whole area to be supported and stating the distance between opposing faces as either not exceeding 2 m, 2 to 4 m, or exceeding 4 M (SMM D20.7). The builder may price the item at a nominal rate to cover a small risk, or at the full value of complete support, or at some intermediate rate proportionate to the amount of support considered necessary. No extra excavation is measured to accommodate earthwork support, the contractor having to make due allowance for this. Earthwork support does not have to be measured for excavations the depth of which is below 250mm.

In Example 1, the earthwork support has been measured to the face of the surface strip, although that face does not exceed 250 mm, because the edge of the strip is in the same plane as the trench under and the total face to be supported exceeds 250 mm. Had the strip for any reason extended beyond the face of the trench excavation then support would not be measured.

DISPOSAL OF EXCAVATED MATERIAL

It must be ascertained how the excavated material is to be disposed of. It is naturally cheaper if the material can be disposed of on the site, but there is often no room for it, and it must then be removed from site. The special circumstances of each case must therefore be considered and the disposal fully described accordingly. Care must be taken to see that every item of excavation in the dimensions has an appropriate item of disposal measured. Where part is to be returned and part removed from site or otherwise disposed of, it will be found simplest to measure an additional item in the first instance as filling with material arising from the excavations equal to the volume of excavation as marked by the cross on Figure 19. When concrete and brickwork are measured later an adjustment can be made of the volume occupied by these as:

Deduct — Filling to excavations average thickness exc. 0.25 m.

&

Add — Disposal of excavated material off-site.

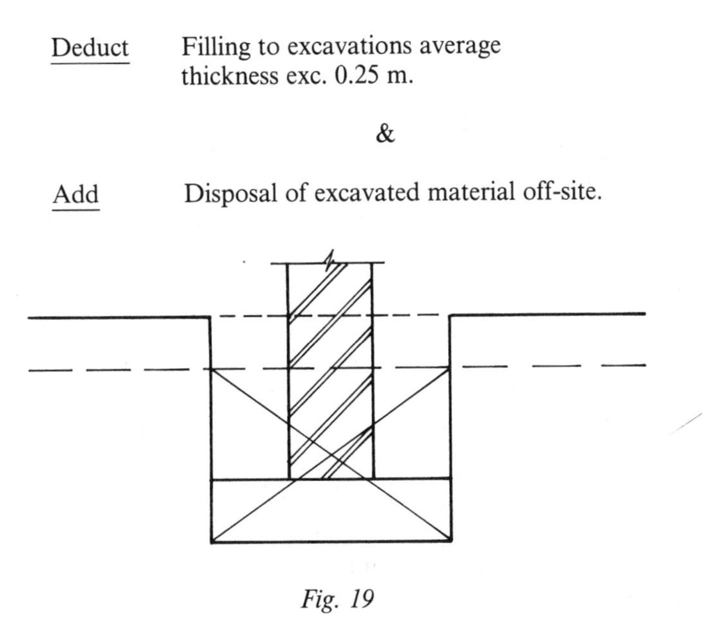

Fig. 19

If filling is deducted for the volume occupied by concrete and brickwork (shown hatched), that remaining of the original measurement will be the volume of space to be filled in. It is simpler to calculate the volume of the brickwork than to arrive at the volume of the spaces on each side. This is one example of the advantage of the over–all system of measurement. Since all is measured in the first instance as filling, if adjustment for removal is forgotten, the error is less serious. Care is necessary in making the adjustment for the volume occupied by the wall to see that the height is not taken above the level to which the filling was measured in the first instance. In the case of basement excavation it would probably be found more convenient to measure all for removal, and adjust subsequently for filling round the outside. If the filling round the walls is to be hardcore then this would be taken in the first instance, instead of filling with material arising from the excavations, together with a removal item for the excavated material. If only the inside of the trench is filled with hardcore then an adjustment will have to be made for the filling to the outside as demonstrated in Example 1.

WORKING SPACE

A superficial item of working space allowance to excavations is taken for formwork, rendering, tanking or protective walls when it is necessary for workmen to operate from the outside and the space available is less than 600 mm. The measurement is taken as the girth or length of the formwork etc. multiplied by the height measured from the commencing level of the excavation to the bottom of the formwork etc. The estimator is left to make a judgement as to the extra work involved.

CONCRETE FOUNDATIONS

The length measured for excavation of trenches will usually be found to serve for the cubic measurement of the concrete. The width will be the full width of trench and the thickness as shown on the drawings. Where the concrete foundation is not at a uniform level throughout it will be found easiest to measure it as if it were, the additional concrete being added afterwards for the laps together with formwork to the face of steps. Again the advantage of over–all measurement shows itself, as if each section were measured piecemeal, there is more danger of error through a section being missed. Concrete poured on or against earth or unblinded hardcore has to be so described.

Concrete foundations are sometimes reinforced by either steel fabric or bars. In such cases it is important to remember that a finer aggregate is necessary than in ordinary mass concrete, and the concrete will therefore be of a different composition but would otherwise be measured in the same way except that it is described as reinforced. A separate linear item is taken for the reinforcing bars which is weighted up in the bill. Fabric reinforcement is measured as a superficial item, stating the width if in a one width strip.

When foundations are reinforced, weak concrete blinding 50 or 75 mm thick is usually required under the foundation. Such blinding is sometimes not shown on the drawings, in which case enquiry should be made of the architect or engineer as to whether it is wanted. Also the engineer may not be satisfied with earthwork support for forming the sides of concrete foundations and will require formwork, and if this is the case, provision for working space will have to be measured. Even if earthwork support is accepted, a flexible barrier or stabilisation of the excavated sides may be required.

The SMM requires that the weight of bars shall include for bends and hooks: for each hook, one can usually add nine times the diameter of the bar rounded to the next 10 mm. If the length of the concrete is used for calculating the length of the bar, then the concrete cover must be deducted, i.e. in the case of a 12 mm bar, 110 mm must be added at each end less the amount of concrete cover.

The text relating to measurement of brick and block walling is left to the next chapter. However, the work in the substructure often forms a separate section in the bill and would therefore be measured in that section as in Examples 1 and 2 that follow. Before working through these examples, Chapter 7 should be read.

Taking–off List	**SMM Reference**
Ground water level	D20.P1(a)
Clearing vegetation etc.	D20.1.3
Excavating topsoil for preservation	D20.2.1
Depositing topsoil	D20.8.3
Excavating to reduce levels	D20.2.2
Disposal of spoil	D20.8.3
Excavating trenches	D20.2.5/6
Filling excavation	D20.9
(If hardcore filling '& on' disposal of spoil)	
Breaking up obstructions	D20.4/5
Work below ground water level	D20.3.1
Compacting bottom of excavation	D20.13.2
Earthwork support	D20.7
Concrete foundations	E10.1
Deduct filling (for volume of foundation)	As above
Add disposal (ditto)	D20.8.3
(Not required if hardcore filling)	
Formwork to foundations	E20.1
Reinforcement to foundations	E30.1/4
Working space allowance	D20.6
Disposal of surface and ground water	D20.8.1/2

For continuation see end of Chapter 7.

EXAMPLE 1
WALL FOUNDATIONS

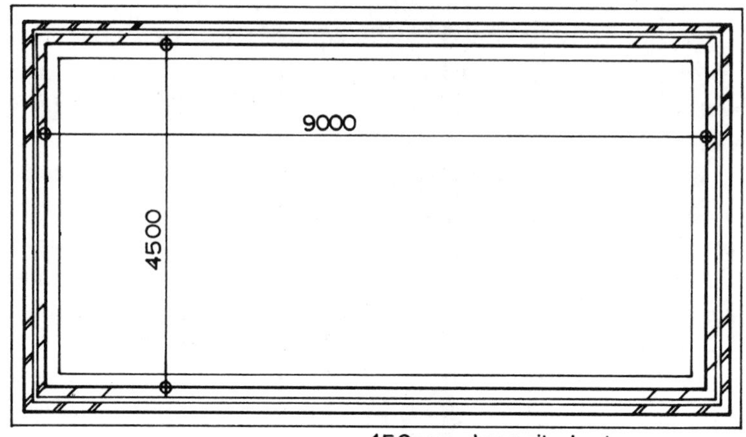

PLAN SCALE 1:100

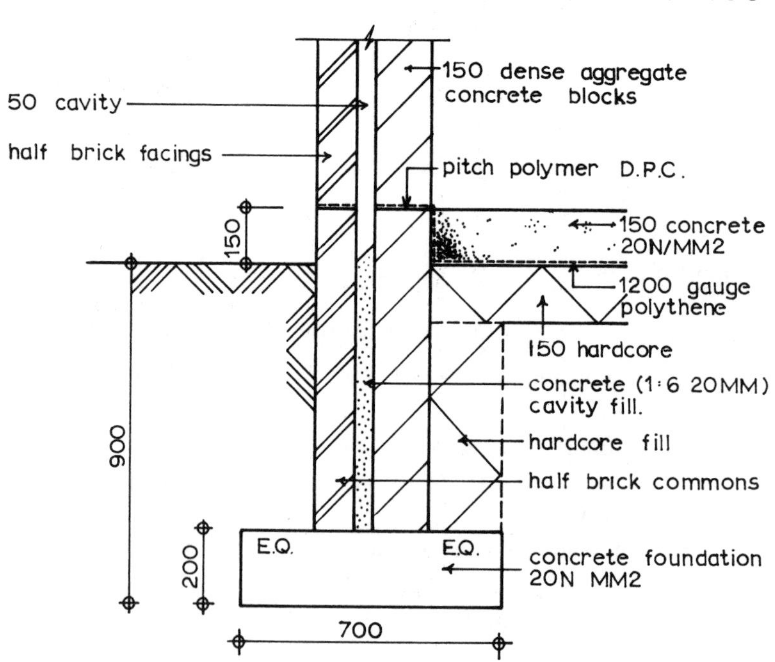

SECTION SCALE 1:20

FIGURE 20

SUBSTRUCTURE 1.

Timesing	Dimensions	Squaring	Description	Notes
			H.B. wall. 102½ Cavity. 50 Inner skin. 150 302½	Calculation of total thickness of wall.
			700 302½ ½/ 397½ Spread of found = 198¾	Calculation of projection of concrete.
			9000 4500 2/302½ 605 605 2/198¾ 397½ 397½ 10 002½ 5502½	Calculation of overall dimensions to outer face of concrete foundation.
	10.00 5.50		Excavating top soil for preservation average 150 mm deep	D20.2.1.1 Top soil excavation only measured when required for re-use.
			&	
			Disposal of excavated material on site in spoil heaps for re-use average 250 m from the excavation x 0.15 = m^3	D20.8.3.2

SUBSTRUCTURE 2.

Mean girths.

	9000 4500	
2/	13500	27 000
4/2/½/	150	600
M/G inner skin.		27 600
	150 50	
4/2/½/	200	800
M/G cavity.		28 400
	102½ 50	
4/2/½/	152½	610
M/G outer skin.		29 010
4/2/½/	102½	410
Outer face wall.		29 420
− 4/2/½/	302½	1210
M/G foundation.		28 210
CHECK −		1210
		27 000

It is advisable to calculate all the mean girths (or centre lines) likely to be required.

Each should be labelled carefully for ease of identification.

Having worked from the inside to the outside it is advisable to go back to the inside measurement as a check.

Depth of trench.

	900
− Topsoil	150
	750

Calculation of trench depth less the topsoil already measured.

SUBSTRUCTURE 3.

Dimensions	Description	Notes
28.21 0.70 0.75	Excavating trenches, width exceeding 0.30m & maximum depth n.e. 1.00m	D20.2.6.2
	&	
	Filling to excavations average thickness exceeding 0.25m with hardcore obtained off-site compacted in 100mm (maximum) layers by vibrator	D20.9.2.3 As one side of the wall is back-filled with hardcore and the other with excavated material a choice has to be made at this stage as to which filling to take here.
	&	
	Disposal of excavated material off-site	D20.8.3.1 If hardcore is chosen for the latter item then the excavated material should be removed from site.
28.21 0.70	Surface treatments compacting bottoms of excavations	D20.13.2.3 Method of compacting would be given in a preamble.

			SUBSTRUCTURE 4.	
			Extl. face wl. 29 420 4/2/198¾ 1 590 31 010	Calculation to find perimeter of external face of trench.
2/	28.21 0.75 31.01 0.15		Earthwork support maximum depth not exceeding 1.00m & distance between opposing faces not exceeding 2.00m	D20.7.1.1 Additional height to outer face
	28.21 0.70 0.20		In-situ concrete (20 N/mm² - 20 mm) foundations poured against earth & Deduct Filling to excavations average thickness exceeding 0.25m with hardcore ab.	E10.1.0.0.5 Hardcore filling is not required for the volume of the concrete.
			900 To DPC 150 1050 Foundation 200 850	Calculation to find height of wall to d.p.c.

SUBSTRUCTURE 5.

	Dimensions		Description	Notes
	29.01 0.85		Walls in commons half brick thickness, vertical, stretcher bond in cement mortar (1:3)	F10.1.1.1 Measurement of outer brick skin from foundation to d.p.c.
	27.60 0.85		Walls in dense aggregate concrete blocks 150mm thickness, in cement mortar (1:3)	F10.1.1.1 Measurement of inner block skin from foundation to d.p.c.
	28.40 0.85		Forming cavities in hollow walls 50mm wide, including galvanised wall ties to BS 1243 type a at the rate of five per m²	F30.1.1.1 Width of cavity is stated
			900 Foundation. 200 700 Splay. 50 750	Calculation to find depth of concrete filling to cavity.

SUBSTRUCTURE 6.

Dims	Description	Notes
28.40 0.05 0.75	In-situ concrete (1:6 - 20mm) filling hollow walls thickness not exceeding 150mm	E10.8.1
	900 Foundation. 200 Topsoil. 150 350 550	
28.21 0.30 0.55	Deduct Filling to excavations average thickness exceeding 0.25m with hardcore a.b.	Deduction of hardcore filling for the volume of the wall up to the top of trench excavation.
	4/2/½/198¾ 29 420 795 30 215	
30.22 0.20 0.55	Deduct ditto & Add Filling to excavations average thickness exceeding 0.25m with excavated material arising from the excavations compacted in maximum 100mm layers & Deduct Disposal of excavated material off-site	Adjustment for earth filling as opposed to hardcore to outside of trench. D20.9.2.1

SUBSTRUCTURE 7.

Dimensions		Description	Notes
30.22 0.02 0.15		Filling to excavations average thickness not exceeding 0.25m with topsoil obtained from on-site spoil heaps average 250m distance, lightly compacted	D20. 9.1.2.3 Replacement of topsoil to outside of wall at ground level.
29.01 0.10 27.60 0.15		Damp proof courses width not exceeding 225mm, horizontal, of Hyload pitch polymer lapped 150mm at joints bedded in cement mortar (1:3)	F30.2.1.3
		[Facings. 3/75 = 225	
29.01 0.23		Deduct Walls in commons half brick thickness a.b. & Add Ditto in facings half brick thickness, vertical, stretcher bond in cement mortar (1:3) including pointing with a weathered joint as the work proceeds one side	F.10.1.1 Facings to exposed external wall taken one course below ground level.

		SUBSTRUCTURE 8.		
			[Ground slab.	
	9.00 4.50 0.15		In-situ concrete ($20 N/mm^2$ - 20mm) beds thickness not exceeding 150 mm	E10.4.1
			&	
			Filling to make up levels average thickness not exceeding 0.25m with hardcore obtained off-site compacted in 100mm (max.) layers by vibrating roller.	D20.10.1.3
	9.00 4.50		Surface treatments compacting and blinding surface of hardcore filling with small stones	D20.13.2.2
			&	
			Damp proof membrane width exceeding 300mm, horizontal, of 1200 gauge polythene sheet to BS 743 laid on blinded hardcore to receive concrete	J40.1.1

SUBSTRUCTURE 9.

			9000 4500 2/ 13500 27 000	
			150 Lap = 50 200	
27.00 0.20			Damp proof membrane width 150-225 mm, vertical, of 1200 gauge polythene sheet to BS 743 between brickwork and edge of concrete slab.	J40.1.1
9.00 4.50			Trowelling on unset concrete bed to receive pavings	E41.3
			– 2/ 198¾ 9000 4500 397½ 397½ 8602½ 4102½	
8.60 4.10			Compacting bottoms of excavations	D20.13.2.3

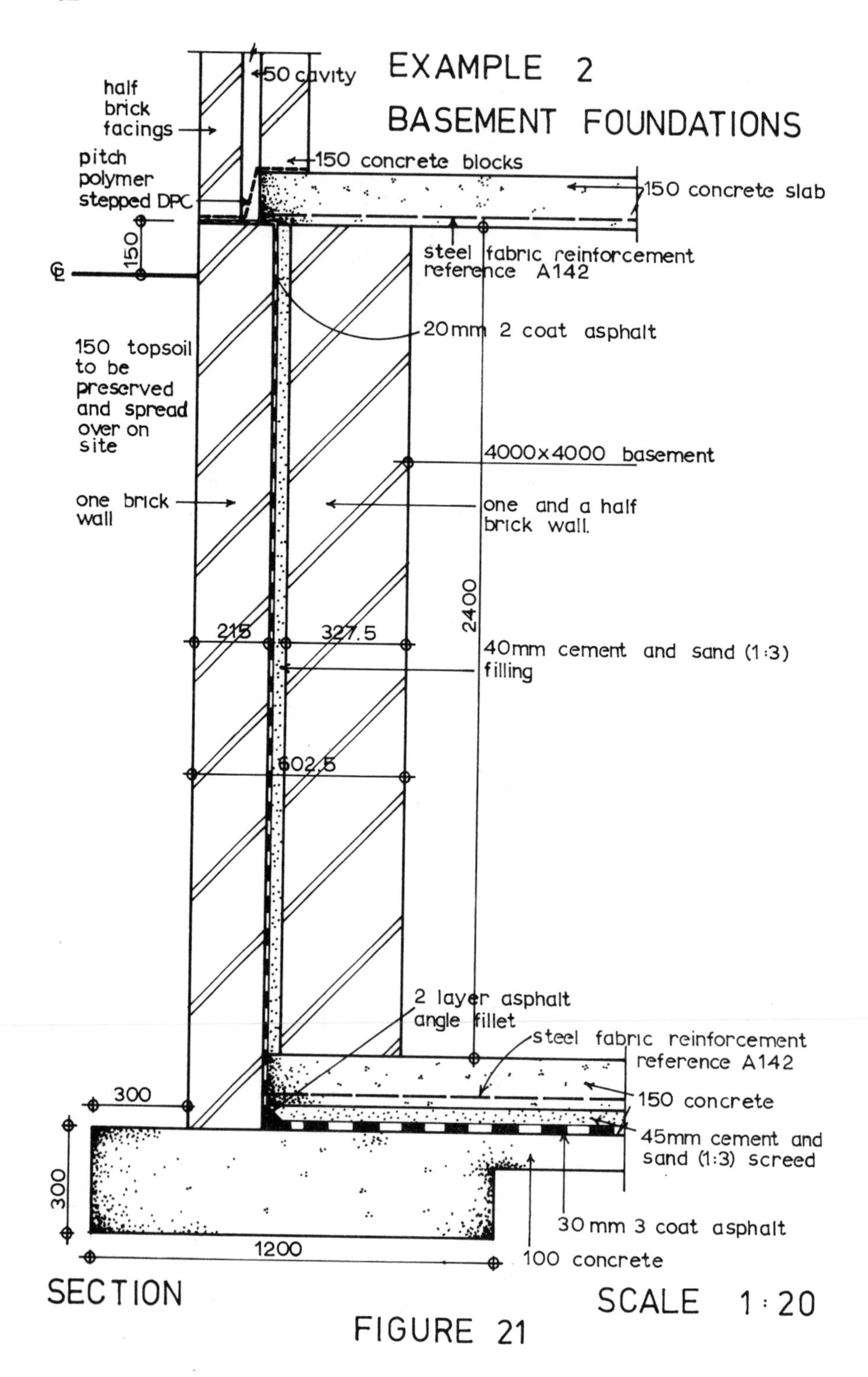
EXAMPLE 2
BASEMENT FOUNDATIONS
half brick facings
50 cavity
150 concrete blocks
pitch polymer stepped DPC
150 concrete slab
150
steel fabric reinforcement reference A142
20mm 2 coat asphalt
150 topsoil to be preserved and spread over on site
4000×4000 basement
one brick wall
one and a half brick wall.
2400
215
327.5
40mm cement and sand (1:3) filling
602.5
2 layer asphalt angle fillet
steel fabric reinforcement reference A142
300
150 concrete
45mm cement and sand (1:3) screed
300
30 mm 3 coat asphalt
1200
100 concrete
SECTION
SCALE 1:20

FIGURE 21

BASEMENT 1.

			Description	Notes
			4000 2/602½ · 1205 2/300 · 600 5805	Calculation of overall dimension to outer face of concrete foundation.
	5.81 5.81		Excavating top soil for preservation average 150mm deep	D20.2.1.1 Topsoil excavation only measured when required for re-use.
			&	
			Disposal of excavated material on-site spread & levelled average 250 m from the excavation	D20.8.3.2
			X 0.15 = m^3	
			2400 Concrete 150 Screed 45 Asphalt 30 Concrete 100 — 325 2725 to DPC 150 top soil 150 — 300 2425	

BASEMENT 2.

Dimensions		Description	Notes
5.81 5.81 2.43		Excavating basements and the like maximum depth not exceeding 4 m	D20.2.3.4
		&	
		Disposal of excavated material off-site	D20.8.3.1 In the case of basements, where most of the excavated material is removed, it is preferable to take this item and adjust for filling later.

GIRTHS.

4/4000	=	16 000
4/2/½/327½	=	1 310
M/G 1B wall	=	17 310
327½ 40		
4/2/½/367½	=	1 470
M/G cav. fill	=	18 780
4/2/½/40	=	160
Face of fill	=	18 940
4/2/20	=	160
Face of asph.	=	19 100
4/2/½/215	=	860
M/G 1B wall	=	19 960
		860
Face brickwk.	=	20 820
4/2/300	=	2 400
Face excavn.	=	23 220
− 4/2/½/1200	=	4 800
M/G foundn.	=	18 420

Calculation of mean girths likely to be required.
Label each girth carefully.

		BASEMENT 3.	
		Check	
		5805 x 4 = 23 220 ✓	Check on face of excavation girth from topsoil dimensions.
		2425 top soil 150 2575	
23.22 2.58		Earthwork support maximum depth not exceeding 4.00m, & distance between opposing faces exceeding 4.00m	D20.7.3.3
		300 Conc. slab 100 200	
18.42 1.20 0.20		Excavating trenches width exceeding 0.30m, maximum depth not exceeding 0.25m commencing 2.58m below ground level	D20.2.6.1.1
		&	
		Disposal of excavated material off-site	D20.8.3.1 As trench is completely filled with concrete disposal, rather than fill, taken at this stage.
18.42 1.20		Surface treatments compacting bottoms of excavations	D20.13.2.3 Method of compacting would be given in a preamble.

		BASEMENT 4.		
	23.22 0.20		Earthwork Support maximum depth not exceeding 4.00m & distance between opposing faces not exceeding 2.00m	D20.7.3.1 Earthwork support not required to inside face of trench as depth does not exceed 0.25 m high.
	18.42 1.20 0.30		In-situ Concrete (20 N/mm² - 20 mm) foundations poured on & against earth	E10.1.0.0.5
			5805 2/1200 = 2400 3405	
	3.41 3.41 0.10		In-situ Concrete (20 N/mm² - 20 mm) beds, thickness not exceeding 150 mm poured on earth	E10.4.1.0.5

BASEMENT 5.

Timesing	Dimensions	Squaring	Description	Reference
	3.41 3.41		Surface treatments Compacting bottoms of excavations	D20.13.2.3
			4000 wall 2/327½ = 655 Fill 2/40 = 80 4735 Asphalt 2/20 = 40 4775	
	4.78 4.78		Trowelling on insect concrete bed to receive asphalt	E41.3
			[Outer wall 2725 Less. Conc. bed = 100 2625 Asphalt DPC = 20 2605	
	19.96 2.61		Walls in commons 1 brick thickness, vertical, English bond in cement mortar (1:3)	F10.1.1.1

BASEMENT 6.

Dimensions	Description	Notes
	[Inner wall	
17.31 2.40	Walls in commons 1½ B thickness ditto	F10.1.1.1
	- 4/2/½/60 : 19 100 240 18 860	
18.86 2.40	Forming cavities in hollow walls 60 mm wide	F30.1.1.1
18.78 0.04 2.40	In-situ cement & Sand (1:3) filling hollow walls thickness not exceeding 150 mm	E10.8.1 Concrete rule used although mortar fill.
	Mastic Asphalt Tanking / Damp proof membranes with limestone aggregate to BS 1097 finished with a wood float using fine sand as abrasive. Subsequently covered	J20.(S1 & S4) Work described as subsequently covered is deemed to include edges and arrises (C3)

BASEMENT 7.

Mastic Asphalt Tanking etc. (Contd.)

Timesing	Dimensions	Squaring	Description	Notes
	4.78 4.78		Damp proofing & tanking level, width exceeding 300mm, 30mm thickness, three coats, laid on concrete	J20.1.4.1 J20 (S2 & S3) Thickness, number of coats and nature of base to be given.
	19.10 2.61		Ditto vertical, width exceeding 300mm, 20mm thickness, two coats to brickwork	J20.1.4.1 Asphalt is measured area in contact with base. Raking out joints brickwork for key deemed included F10.C1(d)
	19.96 0.23		Ditto level, width 150-225mm, 20mm thickness, in two coats to brickwork	J20.1.2.1 Damp proofing to top of 1 B wall.
 4/	19.10 2.61		Solid internal angle fillet to asphalt 45 x 45	J20.12.1 At junction of horizontal & vertical asphalt and at vertical external corners.
	20.82		Fair edges to asphalt	J20.13 To outer edge of d.p.c.

End of Mastic Asphalt

BASEMENT 8.

Dimensions		Description	Ref.
4.74 4.74		Sand - cement (1:3) Screeds, floors, level & to falls only, thickness 45mm in one coat to asphalt base finished with a wood float [Lower Slab.	M.10.5.1.1
4.74 4.74 0.15		In-situ concrete (20 N/mm² - 20 mm) beds thickness not exceeding 150mm, reinforced 4000 2/602½ 1205 5205 Wall 102½ Cavity 50 2/152½ : 305 4900 [Upper Slab.	E.10.4.1.0.1
4.90 4.90 0.15		Ditto slabs do.	E10.5.1.0.1

BASEMENT 9.

Timesing	Dimensions	Squaring	Description	Reference
			Cover 2/20 = 4 735 − 40 = 4 695	
			Cover 2/20 = 4 900 − 40 = 4 860	
	4.70 4.70		Reinforcement for in-situ concrete, fabric to BS 4483 reference A 142 weighing 2.22 kg/m² with 150mm minimum side & end laps	E30.4.1
	4.86 4.86			
			2/150 = 4 900 − 300 = 4 600	
	4.00 4.00		Trowelling on moist concrete bed to receive pavings	E41.3
	4.60 4.60			
			[Upper slab	
	4.00 4.00		Formwork for in-situ concrete soffits of slabs, thickness not exceeding 200mm, horizontal, 1.50-3.00m high	E20.8.1.1.2

BASEMENT 10.

4/	4.90		Formwork for in-situ concrete edges of suspended slabs, plain vertical, height not exceeding 250 mm	E20.3.1.2
			D.P.C. Block 150 Across cavity 160 H.B. 102½ 412½ Face of brick 20 820 102½ 50 150 −4/2/½/302½ = 1210 19 610	
	19.61 0.41		Damp proof courses, width exceeding 225 mm horizontal cavity tray of Hyload pitch polymer bedded in cement mortar (1:3)	F30.2.2.3.1

	BASEMENT II.		
	Backfill.		
	4/2/½/300 = 20 820 1 200 22 020		
	Conc. 2425 100 2325		
22.02 0.30 2.33	Deduct disposal of excavated material off-site & Add Filling to excavations average thickness exceeding 0.25m with excavated material arising from the excavations compacted in maximum 100mm layers	D20.8.3 D20.9.2.1	
22.02 0.30 0.15	Filling to excavations average thickness not exceeding 0.25m with topsoil obtained from on-site spoil heaps average 250m distance, lightly compacted	D20.9.1.2.3 The material required for this filling is deemed to have been taken from an existing spoil heap.	

		BASEMENT 12.		
			[Facework.	
19.96			Deduct brick walling, walls in Commons 1 brick thick a.b.	F10.1.1.1
0.23				
			&	
			Add Brick walling, walls in Commons facework one side 1 brick thick, vertical, English bond, in cement mortar (1:3) including pointing with a weathered joint as the work proceeds one side	F10.1.2

7 Walling

BRICKWORK AND BLOCKWORK IN FOUNDATIONS

This work, as mentioned in the previous chapter, is often kept under a heading of substructure.

Where the concrete foundation is stepped, each length of wall from step to step will have to be measured separately up to damp proof course level. As this is piecemeal measurement the individual lengths taken should be totalled and compared with the mean girth of the entire wall. Alternatively the total length may be measured by the minimum height, the extra heights being added for each section. When measuring internal walls it should be noted that their length will not be equal to that taken for their concrete foundations as the length of these is adjusted at intersections as mentioned in the last chapter.

Brickwork below damp proof course level is commonly specified to be in cement mortar, as opposed to gauged mortar for brickwork above, and this must be made clear in the descriptions. The wall between ground level and damp proof course is often constructed in facing bricks which may be continued for one or two courses below ground level to allow for irregularities in the surface of the ground. The measurement of walling in substructure follows the rules for the measurement of general walling which is covered in the remainder of this chapter.

BRICKWORK THICKNESS

As the thickness of the brickwork has to be given in the description it is usually more convenient to give this in relation to the number of bricks. The generally accepted average size of a brick is 215 × 102.5 × 65 mm, which with a 10 mm mortar joint becomes 225 × 112.5 × 75 mm. This gives the following wall thicknesses:

- half brick 102.5 mm
- one brick 215 mm
- one and a half brick 327.5 mm
- two brick 440 mm

COMMON AND FACING BRICKWORK

When describing brickwork it is necessary to distinguish between common brickwork and facework. Common brickwork is walling in ordinary or 'stock' bricks without any special finish and usually hidden from view. If the brickwork is exposed to view and finished with a neat or fair face then it is known as facework. Often facework is built with a superior type of brick which gives a more pleasing appearance. Walls one brick thick and over may have facework on one side and common bricks on the other, this being a more economical way of constructing the wall if only one face shows. Facework is described as to either one or both sides of the wall.

MEASUREMENT OF BRICKWORK

Brickwork generally is measured as a superficial item taking the mean girth of the wall by the height. The heights of walls often vary within the same building but it is usually possible to measure to some general level, such as the main eaves and then to add for gables etc. and deduct for lower eaves etc. No deductions are made for voids within the area of brickwork not exceeding 0.10 m^2. In the case of internal walls and partitions it is necessary to ascertain whether or not these go through floors and measure the heights accordingly. If the thickness of a wall reduces, say at floor level or at a parapet, then it should be remembered that the mean girth of the wall changes. All openings in walls are usually ignored when measuring at this stage, the deduction for these being made when the doors or windows are measured as described in later chapters.

SUBDIVISIONS OF MEASUREMENT

The measurement of walls can be conveniently divided into the following subdivisions:

(a) external walls
(b) internal walls and partitions
(c) projections of piers and chimney breasts
(d) chimney stacks

The SMM requires that brickwork has to be described in the following classifications:

(1) walls
(2) isolated piers
(3) isolated casings
(4) chimney stacks

In each of these classifications brickwork is described as vertical, battering, tapering (battered one side) or tapering (battered both sides).

Isolated piers are defined as isolated walls when the length of the pier is less than four times its thickness, except where caused by openings.

MEASUREMENT OF PROJECTIONS

Projections are defined as attached piers (if their length on plan is less than four times their thickness) plinths, oversailing courses and other similar items. Projections are measured linear and the width and depth of the projection is given in the description. Horizontal, raking and vertical projections are kept separately.

DESCRIPTIONS

The descriptions for brickwork, in addition to the matters discussed above, have to include the kind, quality and size of the bricks, the bond, the composition and mix of mortar and the type of pointing. These items, particularly if common to all the work, can be included in the bill as preambles or headings to reduce the length of descriptions. It should be remembered that, as already mentioned, mortar mixes may vary and, apart from substructure work, brickwork above eaves level in parapets and chimney stacks may well be specified to be in cement mortar.

CUTTING AND GROOVES ETC.

All rough cutting on common brickwork and fair cutting on facework is deemed to be included. Rough and fair grooves, throats, mortices, chases, rebates, holes, stops and mitres are all also deemed to be included.

RETURNS AND REVEALS

The labour to returns at reveals at the ends of walls is deemed included and therefore nothing has to be measured for these.

HOLLOW WALLS

Each skin of a hollow wall is measured superficially and described as a wall stating whether having facework on one side. A superficial item is measured for forming the cavity stating the width of the cavity. The type and spacing of the wall ties has to be given in the description. If rigid sheet insulation is required in the cavity this is also included in the description stating the type, thickness and fixing method. Foam, fibre or bead cavity filling is measured separately as a superficial item stating the width of the cavity, the type and quality of the material and the method of application. Filling cavities with concrete below ground level is measured as a cubic item stating the thickness in the description as not exceeding 150 mm, 150 to 450 mm or exceeding 450 mm. Closing cavities is measured linear stating the width of the cavity, method of closing and whether horizontal, raking or vertical.

ORNAMENTAL BANDS

These are items such as brick on edge or end bands, basket pattern bands, moulded or splayed plinth cappings, moulded string courses or moulded cornices etc. in facework. Horizontal, raking, vertical and curved bands are measured as separate linear items, stating the width and classified as follows:

(1) flush
(2) sunk (depth of set back stated)
(3) projecting (depth of set forward stated)

If the bands are constructed entirely of stretchers or entirely of headers this has to be stated and if curved, the mean radius given. Ends and angles on the bands are deemed included.

QUOINS

Labours to angles on brickwork are generally deemed included but when quoins, or corners, are formed with facings which differ in kind or size from the general facework they are measured linear on the vertical angle. They are classified in one of the three ways given for ornamental bands above. The method of jointing the quoins to the brickwork has to be stated together with the mean girth.

COPINGS

Copings in facings are measured linear and classified as for ornamental bands above. The description has to give the method of forming and the size. Sills, thresholds and steps are measured in the same way but are more likely to be encountered with the measurement of doors and windows. Tile or slate creasings are also measured linear, the width and number of courses being stated. Ends and angles are deemed to be included.

DAMP PROOF COURSES

These are measured as superficial items and classified by width as not exceeding or exceeding 225 mm. Horizontal, raking, vertical and stepped damp proof courses are each kept separately. Pointing to the exposed edges is deemed to be included but the thickness of the material, the number of layers and the nature of the bedding material has to be stated. In the measurement no allowance is made for laps and no deduction is made for voids, such as flues, not exceeding 0.50 m^2. Asphalt damp proofing and tanking is measured the area in contact with the base as a superficial item and the widths are described as:

(1) not exceeding 150 mm
(2) 150 to 225 mm
(3) 225 to 300 mm
(4) exceeding 300 mm

The thickness, number of coats, nature of base and any surface treatment have to be included in the description together with the pitch. Internal angle fillets are measured as linear items giving a dimensioned description and the number of coats if other than two. Ends and angles are deemed to be included. Fair edges, rounded edges and arrises (or external angles) are measured linear unless the asphalt is subsequently covered when they are included in the description. Raking out joints of brickwork for a key is deemed to be included with the brickwork. No deduction is made for voids in asphalting not exceeding 1 m^2.

REINFORCEMENT

Mesh reinforcement to brick joints is measured linear stating the width and extent of laps. No allowance is made for laps in the measurement.

BLOCKWORK

The rules for measuring blockwork follow very closely those for measuring brickwork. Special blocks used, say, to achieve a designed bond, at reveals, intersections and angles are measured as linear items and described as extra over the work on which they occur.

RENDERING

Rendering to external walls is usually taken with the walls and is measured the area in contact with the base as a superficial item. The description has to include the thickness, number of coats, mix, surface treatment and the work has to be described as external. Rendering not exceeding 300 mm in width is measured as a linear item and work to isolated columns is kept separately. Rounded internal and external angles are measured as linear items. If metal beads are specified then these are measured linear and include the working of finish up to them. Painting to the walls is measured as a superficial item and described as to general surfaces externally.

STONEWORK

The measurement of stone walling is considered to be beyond the scope of this book but generally the rules follow those given for brickwork. Among the exceptions are isolated columns, rough and fair raking and circular cutting, grooves, throats, flutes, rebates and chases all of which are measured linear.

TIMBER AND PATENT PARTITIONS

These partitions are usually measured with this section so as to complete the measurement of the internal walls. When measuring timber or stud partitions all timbers including struts and noggings are measured linear and described as wall or partition members giving their size in the description. All labours in the construction of the partition are deemed to be included. Door openings are usually measured net, as an adjustment later with the doors could be rather complicated. Plasterboard and other finish are usually measured with the internal finishings section. Patent partitions are measured their mean length as linear items stating their height in 300 mm stages and the thickness. All sole and head plates, studs etc. are deemed to be included. Openings in the partitions are ignored unless they are full height in which case they are deducted from the length. Angles, tee junctions, crosses, fair ends and abutments to other materials are measured as linear items.

Taking-off List	**SMM Reference**
Substructure:	
Brick or block external walls	F10.1
Forming cavities in hollow walls	F30.1
Filling hollow walls with concrete	E10.8
Brick or block internal walls	F10.1
Projections to walls	F10.5
Deduct filling to excavations	D20.9
Add disposal of excavated material	D20.8.3
Topsoil adjustment to outside of wall	D20.9
Damp proof courses	F30.2
Asphalt tanking	J20.1
Internal angle fillets to tanking	J20.12
Edges and arrises to tanking	J20.13/16
Wall facework at ground level	F10.1
Concrete bed	E10.4
Hardcore bed	D20.10
Compacting and blinding hardcore	D20.13.2.2
Compacting ground	D20.13.2.1/3
Damp proof membrane	J40.1
Vertical ditto	J40.10
Trowelling concrete	E41.3
Insulation	P10.1
Superstructure:	
Brick or block external walls	F10.1
Forming cavities in hollow walls (including rigid insulation)	F30.1
Cavity insulation (foamed, fibre and bead)	P11.1
Brick or block internal walls	F10.1
Projections to walls	F10.5
Chimney stacks	F10.4
Ornamental bands	F10.13
Quoins	F10.14
Brick copings	F10.17
Damp proof courses	F30.2
External rendering	M20.1
Rounded angles to ditto (10–100 mm radius)	M20.16
Angle beads etc.	M20.24.8
Decoration to rendering	M60.1
Timber partitions	G20.7
Patent partitions	K31.1

EXAMPLE 3

WALLS AND PARTITIONS (above damp proof course)

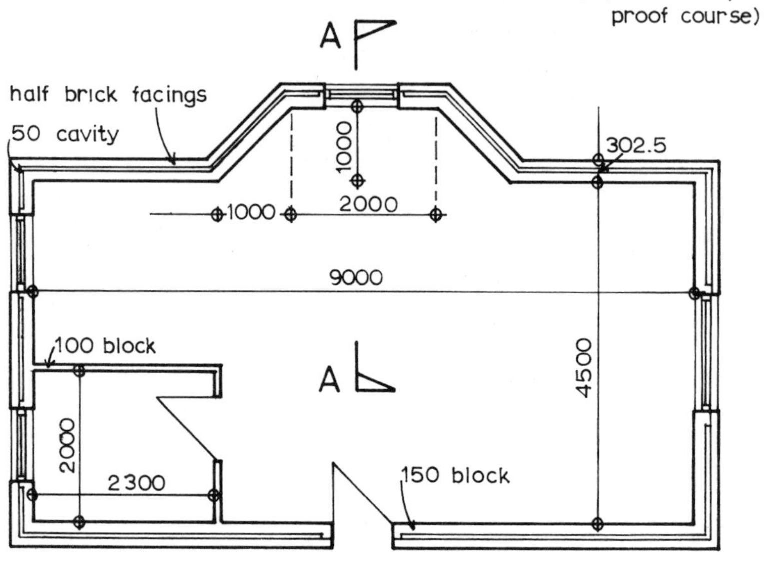

PLAN

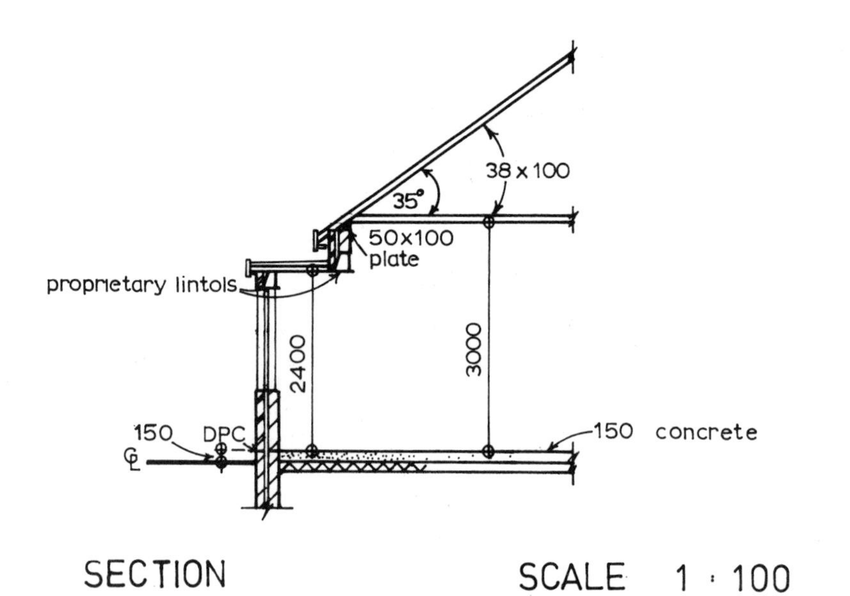

SECTION SCALE 1 : 100

FIGURE 22

EXTERNAL WALLS 1.

[Calculation of mean girths excluding bay.

9000
4500
2/13500

Internal face = 27000

4/2/½/150 = 600

M/G int. skin = 27600

150
50
4/2/½/200 = 800

M/G cavity = 28400

102½
50
4/2/½/152½ = 610

M/G Ext. skin = 29010

4/2/½/102½ = 410

Ext. face = 29420

As the wall to the projecting bay is a different height it has been decided to ignore this in the first instance and make an adjustment later.

CHECK.

9000 4500

150
50
102½
2/302½ = 605 605

9605 5105
9605
2/14710
= 29420 ✓

It is always advisable to check the mean girth by repeating the calculations in a different way. Here the external face calculated above is checked by adding the overall dimensions.

Floor to clg. 3000
- Plate 50
2950

EXTERNAL WALLS 2.

	Dimensions	Squaring	Description	Reference
			[Int. skin.	
	27.60 2.95		Walls in dense aggregate concrete blocks, 150mm thickness, vertical, in gauged mortar (1:1:6)	F10.1.1.1
			[Outer skin.	
	29.01 2.95		Walls in facings, half brick thickness, vertical, stretcher bond in gauged mortar (1:1:6) including pointing with a weathered joint as the work proceeds one side	F10.1.1.1
			[Cavity.	
	28.40 2.95		Forming cavities in hollow walls 50mm wide including galvanised wall ties as BS 1243 type a at the rate of five per m^2	F30.1.1

EXTERNAL WALLS 3.

		Description	Notes
		Adjustment for bay.	
		Difference in length on internal face.	Sketch to show that difference remains constant.
		Bay 2000 2/1000 √2 = 2828 4828 Deduct 2000 2/1000 2000 4000 Additional length = 828	Length x remains the same for both walls. Length y increases by the same amount as z decreases.
0.83	2.40	Block walling, 150 mm thick a.b. & Walls in facings half brick thickness a.b. & Forming cavities in hollow walls a.b.	In practice this dimension could be bracketed on to each of the last three dimensions but for clarity is kept separately here.

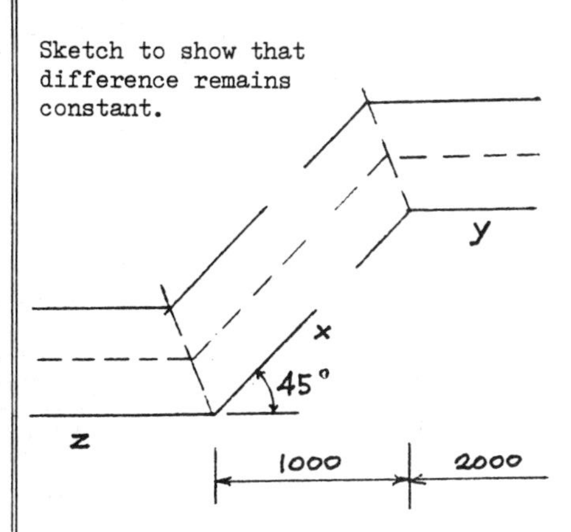

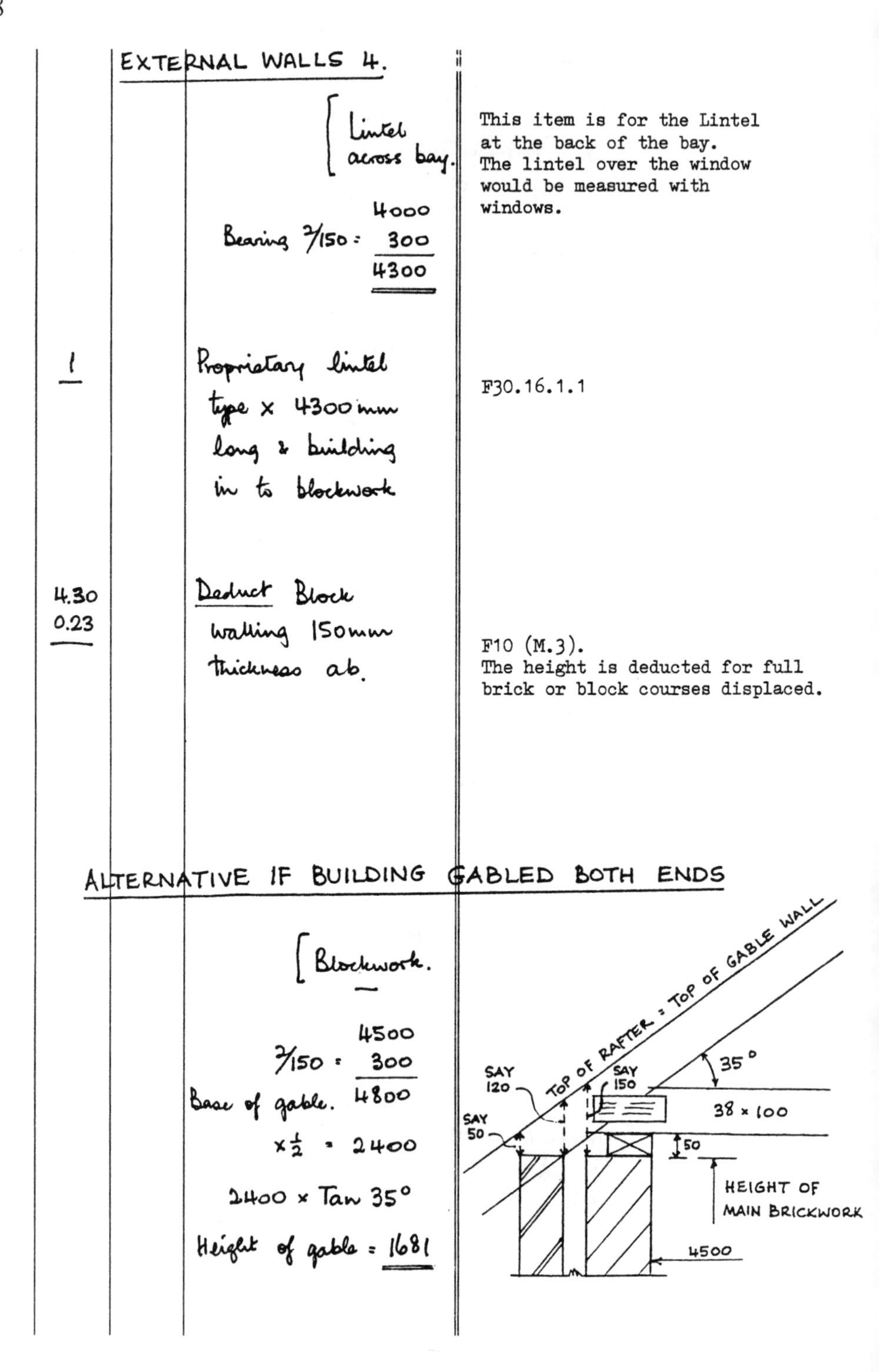
EXTERNAL WALLS 4.
Lintel across bay.
This item is for the Lintel at the back of the bay. The lintel over the window would be measured with windows.
4000
Bearing 2/150 = 300
4300
1
Proprietary lintel type x 4300mm long & building in to blockwork
F30.16.1.1
4.30
0.23
Deduct Block walling 150mm thickness ab.
F10 (M.3).
The height is deducted for full brick or block courses displaced.
ALTERNATIVE IF BUILDING GABLED BOTH ENDS
Blockwork.
4500
2/150 = 300
Base of gable. 4800
x ½ = 2400
2400 x Tan 35°
Height of gable = 1681
TOP OF RAFTER = TOP OF GABLE WALL
SAY 120
SAY 150
SAY 50
35°
38 x 100
50
HEIGHT OF MAIN BRICKWORK
4500

EXTERNAL WALLS 5.

2/	4.80 0.15		Block walling 150mm thickness a.b.
2/½/	4.80 1.68		

In practice these dimensions could be bracketed on to the first three items but for clarity they are kept separately here.

[Brickwork.

4500
2/302½ = 605
Base of gable. 5105
× ½ = 2552

2552 × Tan 35°

Height of gable. = 1786

2/	5.11 0.05		Walls in facings half brick thickness a.b.
2/½/	5.11 1.79		

EXTERNAL WALLS 6.

[Cavity.

			4500
		150	
		50	
		2/ 200 =	400
			4900
		× ½ =	2450

2450 × Tan 35°

Height of gable = 1716

Timesing	Dimensions	Squaring	Description
2/	4.90 0.12		Forming cavities in hollow walls a.b.
2/ ½/	4.90 1.72		

INTERNAL WALLS 1.

		2300
		2000
		4300
		2/½/100 = 100
		4400
4.40 3.00	Walls in dense aggregate concrete blocks 100mm thickness, vertical, in gauged mortar (1:1:6)	F10.1.1.1

ALTERNATIVE FOR STUD PARTITION.

Assume 100 x 50 head, cill & studs at 450 c/cs.

2300
100
450) 2400 (5.4
Say 6 + 1
= 7

450) 2000 (4.7
Say 5 + 1
= 6

+1 Extra stud at Corner.

7 + 6 + 1 = 14

INTERNAL WALLS 2.

Timesing	Dimensions	Description	Side notes
		3000 Head/cill 2/50 = 100 2900	
2/	2.40	Wall or partition members 100 x 50mm sawn treated softwood	G20.7.0.1
2/	2.00		Note partitions include struts and noggings.
14/	2.90		
		Noggings. 2400 2000 4400 14/50 = 700 3700	The method of fixing need not be included if at the discretion of the contractor. G20 (S2) However fixing through vulnerable finishes such as glass must be so described G20 (S3)
	3.70		
		[Adjustment for door. [Assume size 900 x 2100	It is usual to measure stud partitions net.
		2100 50 2150	
	2.15	Deduct	One stud displaced by door.
	0.85	900 50 850 [Nogging.	
	0.90	Add [Head	

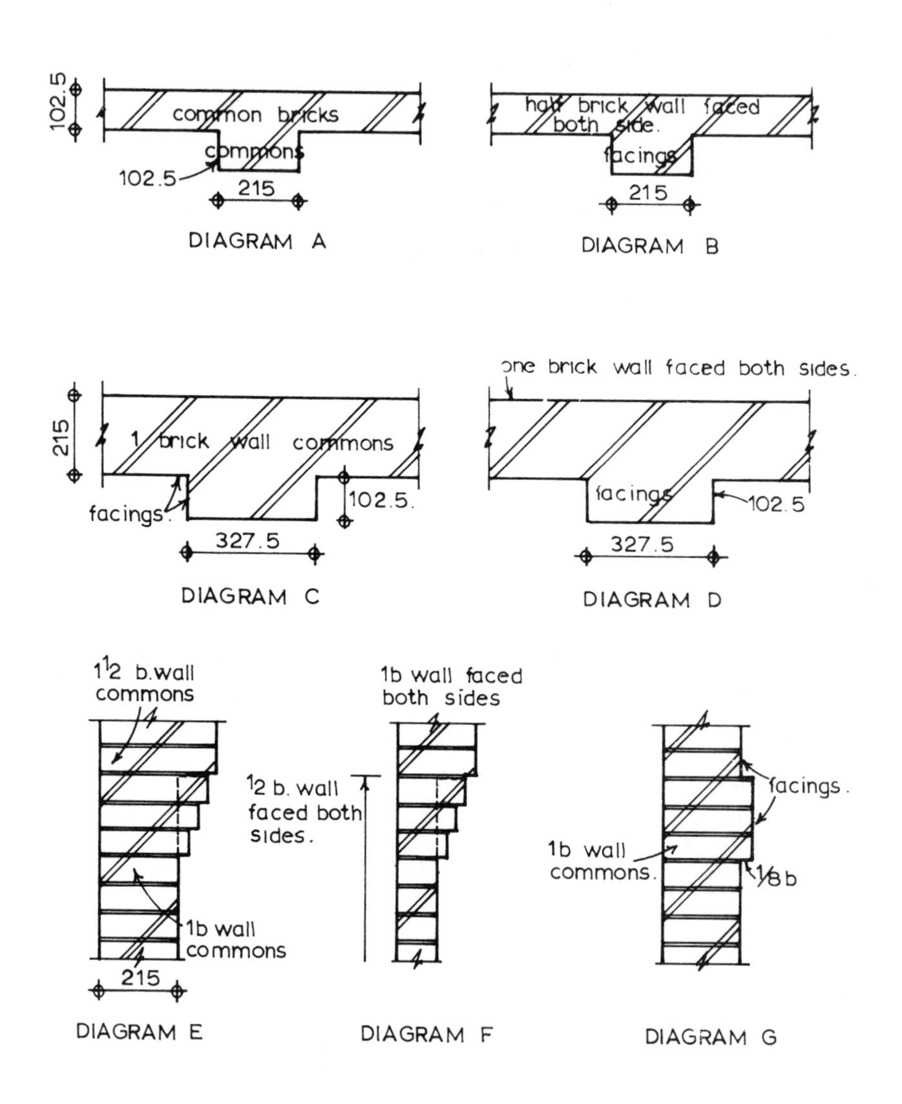

PLANS DIAGRAMS A B C & D ALL 3000 HIGH
SECTIONS DIAGRAMS E F G ALL 5000 LONG

FIGURE 23

BRICKWORK 1.

Note: In the following examples it has been assumed that the main wall has been measured ignoring the projection.

DIAGRAM A

3.00	Projections in common bricks, one brick wide and half brick projection, vertical, in gauged mortar (1:1:6)	F10.5.1.1 Measured as a projection if the length of the pier on plan does not exceed four times the projection.

BRICKWORK 2.

DIAGRAM B

0.23 3.00	Deduct Walls in facing bricks, half brick thickness, vertical, pointed both sides ab. & Add Ditto but pointed one side	The original half brick wall in facings pointed both sides is substituted, for the area of the pier, with a similar wall pointed one side.
3.00	Projections in facing bricks, one brick wide and half brick projection, vertical, in gauged mortar (1:1:6) including pointing with a weathered joint as the work proceeds	F10.5.1.1 F10.C1 (f) states that labours on returns are deemed included and therefore there is no requirement to refer to the pointing necessary to the two returns.

		BRICKWORK 3.	
		DIAGRAM C	
0.33 3.00		Deduct Wall in common bricks, one brick thickness, vertical, faced and pointed one side a.b. & Add Wall in common bricks, one brick thickness, vertical, English bond, in gauged mortar (1:1:6)	The original one brick wall in commons faced and pointed on one side is substituted, for the area of the pier, with a one brick wall in commons.
3.00		Projections in facing bricks, in English bond, one & a half bricks wide and half brick projection, vertical, in gauged mortar (1:1:6) including pointing with a weathered joint as the work proceeds	F10.5.1.1

			BRICKWORK 4.	
			DIAGRAM D	
	0.33 3.00		Deduct Walls in facing bricks, one brick thickness, vertical, pointed both sides a.b.	The original one brick wall in facings pointed both sides is substituted, for the area of the pier, with a one brick wall in commons faced and pointed on one side.
			&	
			Add Wall in Common bricks, one brick thickness, vertical, English bond, in gauged mortar (1:1:6) faced and pointed with a weathered joint as the work proceeds one side	
	3.00		Projection in facing bricks, in English bond, one & a half bricks wide & half brick projection, vertical, in gauged mortar (1:1:6) including pointing with a weathered joint as the work proceeds	F10.5.1.1

BRICKWORK 5.

DIAGRAM E

		Bottom course $\frac{1}{8}$B Top " $\frac{3}{8}$B $\div 2/$ $\frac{1}{2}$B = $\frac{1}{4}$B avge	Calculation to find the average thickness of the three course projection.
5.00		Projections in common bricks, average quarter brick thick and three courses high, horizontal, in gauged mortar (1:1:6)	F10.5.1.3

BRICKWORK 6.

DIAGRAM F

3/75 = 225

5.00 0.23	Deduct Walls in facing bricks, half brick thickness, vertical, pointed both sides a.b. & Add Ditto but pointed one side	The original half brick wall in facings pointed both sides is substituted, for the area of the oversailing courses, with a similar wall pointed one side.
5.00	Projections in facing bricks average quarter brick thick and three courses high, horizontal, in gauged mortar (1:1:6) including pointing with a weathered joint as the work proceeds	F10.5.1.3

BRICKWORK 7.

DIAGRAM G

5.00 0.23	Deduct Wall in Common bricks, one brick thickness, vertical, faced and pointed one side ab. & Add Wall in common bricks, one brick thickness, vertical, English bond, in gauged mortar (1:1:6)	The original one brick wall in commons faced and pointed one side is substituted, for the area of the projection, with a one brick wall in commons.
5.00	Projections in facing bricks, one eighth brick thick and three courses high, horizontal, in gauged mortar (1:1:6) including pointing with a weathered joint as the work proceeds	F10.5.1.1

8 Fires and vents

Careful consideration has been given by the authors as to whether or not to retain this chapter. As this book deals mainly with domestic construction it was considered that the revival of popularity for fireplaces in living rooms warranted an inclusion of some written material on this section of work.

There are a considerable number of building regulations applying to hearths, fireplace recesses, chimneys and flues and the taker–off should be familiar with these before measuring any part of this work. Generally, additional foundation work, chimney breasts, damp proof courses and stacks will have been measured with the previous sections of substructure and brickwork. The creation of openings in floors and roofs is usually left to be dealt with in these respective sections of work and is discussed later. Also additional finishings to the chimney breast is more conveniently taken with internal finishings. Any boilers for central heating or hot water systems with their flues are generally measured with the heating and hot water installation but fireplaces with back boilers may be taken in this section. As with all sections of work where the divisions are not clearly defined there should be discussion between takers–off to ensure that work is neither duplicated nor omitted altogether.

SUBDIVISIONS

The measurement of fires sub–divides itself into four main sections:

(1) adjustment for fire opening including lintel etc.
(2) flue liners, chimney caps and pots
(3) structural hearths etc.
(4) fire surround, hearth, bottom grate and fret

ADJUSTMENT FOR FIRE OPENING

The brickwork to the chimney breast will usually have been measured solid and it is necessary at this stage to create an opening for the fire although freestanding appliances may not require a recess. Care must be taken that the thickness of the projecting jambs and the wall at the back of the opening comply with the appropriate regulations.

The projecting brick chimney breast may have been measured as a linear item stating the width and depth of the projection. If, however, the width of the projection of the chimney breast exceeds four times its depth or projection then the chimney breast including the wall at the back will have been taken as solid brickwork, that is, measured superficially stating the thickness. Whatever has been measured, a deduction will have to be made for the height of the opening and the brick jambs on either side of the opening measured linear as projections.

The support of the brickwork above the opening is probably best achieved with a precast concrete throat unit which also reduces the size of the top of the recess to meet the flue. Alternatively the traditional method is to use a splayed lintel at the front, corbel the brickwork to reduce the size of the opening and render the corbelling to make a smooth junction with the flue liner. Rendering may be required to the outside face of the chimney breast where it passes through timber floors and roofs.

FLUE LINERS ETC.

For domestic work flue liners are usually a minimum of 200 mm diameter or 200 mm square, the latter being more satisfactory for building into masonry chimneys. Flue liners are measured linear and the description is deemed to include, for cutting to form easings and bends and cutting to walls around. No deduction of brickwork is made for flues and liners if the area displaced by both is less than 0.25 m^2. Chimney pots and precast concrete chimney caps are enumerated.

Typical descriptions for these items are as follows:

Linear	212 × 212 mm flue lining of clayware as BS 1181 Type A1 square section with rebated joints, bedded in fireclay.
Number	600 mm high chimney pot as BS 1181 type D with 212 × 212 mm square base set and flaunched in cement mortar (1:3).

Number 1000 × 775 × 225 mm thick precast concrete (mix) chimney cap with weathered top, with two 225 × 225 mm perforations, throated all round, finished fair on exposed faces and bedded in cement mortar (1:3).

HEARTHS

The structural hearth in a solid ground floor is usually formed with the concrete ground slab and an adjustment to the measurement of this may have to be made. Hearths in suspended timber ground and first floors are usually constructed of solid concrete and are measured at this stage. In situ concrete hearths are measured as cubic items and described as slabs classifying the thickness in the description. Formwork and reinforcement are measured separately. Precast concrete slabs are enumerated stating the size, reinforcement being included in the description. Fender walls, damp proof courses and hardcore fill to suspended timber ground floors must not be forgotten. If an underfloor air supply and sunken ashpit is required, the pipe and work forming the pit would be measured at this stage.

FIRE SURROUNDS AND HEARTHS

These may consist of faced precast slabs the vertical part being fixed to the wall with metal lugs. Commonly a PC or provisional sum is included for the supply of this item and the fixing and bedding is measured. A brick or stone fireplace surround built in situ is usually constructed by a specialist in this type of work and would again be included as a PC or provisional sum. The refractory concrete fireback is enumerated describing the item including its bedding. The fire bottom and fret would probably be included in the PC or provisional sum and, if for supply only, an item taken for placing these in position. An adjustment of floor and wall finishings and decorations displaced by the surround and hearth could be made either at this stage or, more satisfactorily, later with the internal finishings section.

FLUE PIPES

These are usually associated with boilers and would therefore be taken with the heating and hot water installation.

CONCRETE FLUE BLOCKS

Gas flue blocks are enumerated giving their size and number of flues in the description of each block.

AIR BRICKS AND SOOT DOORS

Air bricks are traditionally measured at this stage and possibly soot doors unless they are related to the heating installation. Air bricks used for roof and underfloor ventilation may be taken with their respective sections of work. Air bricks, ventilating gratings and soot doors are enumerated. The description, apart from the item itself, must give the nature and thickness of the wall and details of any arches or lintels. The actual formation of the opening is deemed included.

Taking-off List

It is assumed that the extra foundation, chimney breast and stack have been measured and that the creation of the floor and roof openings and extra finishings to the chimney breast are measured elsewhere.

Taking-off List	SMM Reference
Adjustment of fireplace opening	As before
Lintel or throat unit	F31.1
Rendering to top of fire opening	M20.1
Rendering to outside of chimney breast	M20.1
Flue linings	F30.11
Chimney pot	F30.16
Chimney cap (precast concrete)	F31.1
Concrete hearths (precast)	E50.1
Ditto (in situ)	E10.5
Formwork	E20.8/3
Reinforcement	E30.1–4
Fender walls	F10.1
Damp proof course	F30.2
Hardcore, compacting and blinding	D20.10/13
Ventilation pipes	Y10.1
Sunken ashpit	F30.16
Fireback	F30.16
Fireplace surround and hearth	N10.1
Adjustment of finishings for ditto	As finishings
Fire bottom and fret	F30.16
Soot doors	F30.14
Air bricks	F30.12
Gas flue blocks	F30.15

9 Floors

TIMBER SIZES

Timber for use in construction is sawn into standard sectional dimensions and the size thus created is known as the nominal or basic size. When the timber is processed, planed or wrot it is reduced in size, usually by about 2 mm on each face and the resultant size is known as the finished size. The original size of processed timber is sometimes described as 'ex' or 'out of', thus 'ex 100 × 25 mm' would be '96 × 21 mm finished'.

Regularising is a machine process by which structural timber is sawn to a uniform size e.g. joists sawn to an even depth. For regularising 3 mm should be allowed off the size of timber up to 150 mm and 5 mm above this size.

The SMM states that sizes of timber are deemed to be nominal unless described as finished. Confusion may be avoided if the sizes given in the bill are as they are shown on the drawings provided that these are consistent. Structural timber when not exposed is usually left with a sawn finish and the sizes given on the drawing are nominal. When architects are designing joinery, which may have to be to precise dimensions, they often prefer to give finished sizes. Care must be taken to ascertain which sizes are given on the drawing and to make it abundantly clear in the bill if finished sizes are being used.

The amount of timber removed in creating a wrot face on timber is known as the planing margin and this should be given as a preamble in the bill. Reference may be made to BS 4471 'Dimensions for Softwood', which sets out in detail various planing margins according to the sizes and the 'end use' of the timber.

The specification for tongued and grooved floor boarding can cause confusion as the finished width on face and the finished thickness may be quoted e.g. 90 × 21 finished boarding would be 100 × 25 nominal (sawn) size to allow for the tongue and planing margin.

Further consideration should be given to the sizes of timber readily available. For example the width of a door lining to a 75 mm block partition with 13 mm plaster on both faces would require a finished lining width of 101 mm. The nearest nominal size of timber available is 115 mm or even 125 mm for some types of timber.

Timber exceeding 5.1 m in length is usually more expensive and in the examples an allowance in the length of timbers has been made for jointing where appropriate. The SMM requires structural timbers over 6 m long in one continuous length to be measured as such including stating the length SMM (G20.6.0.1.1).

SUBDIVISIONS OF MEASUREMENT

The measurement of floors is conveniently divided into two main subdivisions of finishings and construction. These subdivisions may be taken storey by storey or the finishings taken first to the whole building followed by the construction. Whether or not the finishings are taken before the construction is really a matter of personal choice although measurement of, and therefore knowledge of, the finishings may assist in deciding the form of construction to be used. For example a timber upper floor may have a finish of boarding generally but a small tiled area may require a different form of construction.

Frequently floor finishings are measured with the internal finishings section as measurements used for ceilings may apply to floors. This is, however, a matter for agreement between the surveyors measuring the two sections although it is customary to measure finish to timber floors with the construction.

MEASUREMENT OF FLOOR FINISHINGS

Both in situ and tile finishings have to be described as follows:

(1) level or to falls only less than 15 degrees
(2) to falls and crossfalls less than 15 degrees
(3) to slopes more than 15 degrees

For in situ, sheet and flexible tile floorings the area measured is that in contact with the base. For rigid tile flooring the finished face is measured although in practice, as far as floors are concerned, this usually makes little difference. No deduction is made for voids not exceeding 0.5 m^2 within the areas of the work. If the work is laid in bays this has to be stated giving their average size. Work in staircase areas and plant rooms has to be so described.

Skirtings are usually measured with internal finishings as their length relates to that of the walls. Floor coverings in door openings are either taken with doors, as the surveyor dealing with these is familiar with the dimensions, or with the finishings. It should be remembered that dividing strips may be required where the finish changes. Doors are usually hung flush with the face of the wall of the room into which they open; therefore the floor finish in the opening should be that of the room which shows when the door is closed. Mat frames and mat wells should be measured with floor finishings.

Screeds, if required, are measured with floor finishings as their areas are usually the same. Care should be taken to ascertain the thickness of screeds where the floor finishings vary. Usually it is necessary to have floor surfaces level and differing finishings may have varying thicknesses requiring the screed to take up the difference. Sometimes this difference is allowed for in the floor construction.

Timber boarding, plywood or chipboard to floors is measured superficial unless it is not exceeding 300 mm wide when it is measured linear. Areas not exceeding 1 m^2 are enumerated. No deduction is made for voids not exceeding 0.5 m^2 within the area of the flooring. Remember that nosings and margins may be required around openings in the floor.

TIMBER FLOOR CONSTRUCTION

The measurement of timber floors may be divided as follows:

- plates and/or beams
- joists
- hangers
- strutting
- ties to walls
- insulation

Timber suspended ground floors are not usually cost effective but may be used to avoid excessive fill below solid floors or on sloping ground. The under floor space should be ventilated and air bricks and sleeper walls built honeycombed with dpc's may be required and are either measured with floors or substructure but would be billed with the latter.

Timber plates and bearers are measured linear and the length measured should allow for any laps or joints required. Timber joists are measured in the same way but are described as floor members. The length taken for the joists should include for building in if required. If preservative treatment is specified to the ends of joists this is enumerated but general preservative treatment to timber would be described with the timber or as a preamble.

To ascertain the number of joists in a room, take the length of the room at right angles to the span of the joists and subtract, say, a 25 mm clearance plus half the thickness of the joist at each end. This will give the distance from the centre lines of the first and last joints which is divided by the spacing of the joists. The result, which should be rounded to the next whole number, gives the number of spaces between the joists. To this number one must be added to give the number of joists rather than the spaces. Two points should be borne in mind before making the calculation. Firstly the room size used for the calculation should be that of the room below the floor as the joists will be supported by these walls. Secondly, if any of the intermediate joists are to be in a fixed position, such as a trimmer, two separate dimensions should be taken for the division on either side of the fixed joist, thus avoiding any increase in the maximum spacing. Generally for the entire building it is more satisfactory to run all the joists in the same direction and to keep them all the same depth. This avoids a change in direction of boarding, if used, and a change in ceiling level to the rooms below. Additional joists may be required to support upper floor partitions and care must be taken to include for these.

The trimming of timber joists for staircases, ducts, hearths, trap doors, etc. is usually taken with the floor construction. Joists used in trimming should be increased in thickness by 25 mm and an addition made to their length for jointing unless metal hangers and framing brackets are used. Displaced joists and floor coverings have to be deducted.

To prevent joists turning herringbone or solid strutting, which may not be shown on the drawings should be taken. This is measured linear over the joists and there should be one row, say, every 2.4 m.

Building regulations may require ties to be provided where timber joists run parallel to an envelope wall and these are probably best taken with the floor construction.

CONCRETE FLOORS

Concrete ground slabs laid on hardcore frequently form the ground floor. These are sometimes measured with foundations as any work up to damp proof course level is included in the substructure section of the bill. The measurement of these slabs has been covered in Example 1.

Suspended in situ concrete floors may be measured in three main sections:

- slab and beams
- formwork
- reinforcement

The concrete slab is most conveniently measured over the beams, not forgetting that the bearings in walls must be included in the volume. An indication of the thickness of the slab has to be given in the description as: not exceeding 150, 150 to 450 and exceeding 450 mm. Beams are measured as cubic items and, unless they are deep, are added to the volume of the slab.

Formwork is measured to the area of concrete to be supported which, of course, excludes the area of bearings. The slab thickness is classified in the description as not exceeding 200, 200 to 300 and then in 100 mm stages. The propping height is required to be given in 1.5 m stages. Formwork to edges of slabs is measured linear and described as not exceeding 250 mm and 250 to 500 mm deep. Formwork to regularly shaped beams is measured superficial stating the number of beams in the description. If the formwork to the slab has been measured across the beams then this is deducted but no deduction is made where beams intersect each other.

Bar reinforcement is measured linear adding the length of hooks, anchorages, cranks and laps; the linear measurement being converted to tonnes in the bill. Given the spacing of the bars, the number may be calculated in a similar manner to floor joists. Fabric reinforcement is measured superficial, the net area being taken excluding laps.

Taking-off List	**SMM Reference**
Suspended ground floor:	
It is assumed that the hardcore, concrete bed and damp proof membrane have been measured with the substructure.	
Sleeper walls	F10.1
Damp proof course	F30.2
Air bricks	F30.12
Walls plates	G20.8
Joists	G20.6
Insulation	P10.2
Boarding	K20.2
Timber upper floor:	
Wall plates	G20.8
Timber beams and joists	G20.6
Joist hangers and connectors	G20.21/24
Strutting	G20.10
Ties to walls	G20.20
Insulation	P10.2
Boarding	K20.2
Opening adjustments	As before
Nosings and margins	P20.2
Solid ground floor:	
It is assumed that the hardcore, concrete bed, damp proof membrane and insulation have been measured with the substructure.	
Floor finish (may be taken with internal finishings)	M10/12.5 M40/42.5 M50.5
Screeds (ditto)	M10.5
Dividing strips (ditto)	M12.24.7 etc
Concrete upper floor:	
Floor finish, screeds and dividing strips	as above
Concrete slab	E10.5
Concrete beams	E10.9
Formwork	E20.8/13
Reinforcement	E30.1–4
Trowelling to slab	E41.3

EXAMPLE 5

SUSPENDED TIMBER FLOOR

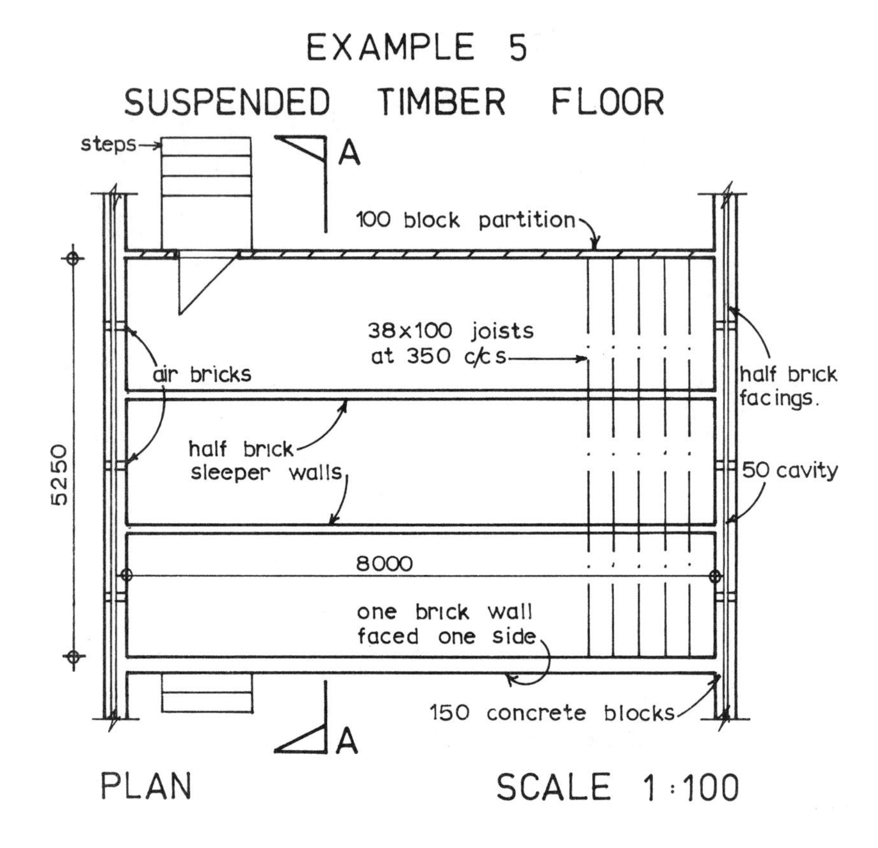

NOTE: Steps, partitions, external walls, finishings, ground slab, hardcore and flooring in opening measured elsewhere.

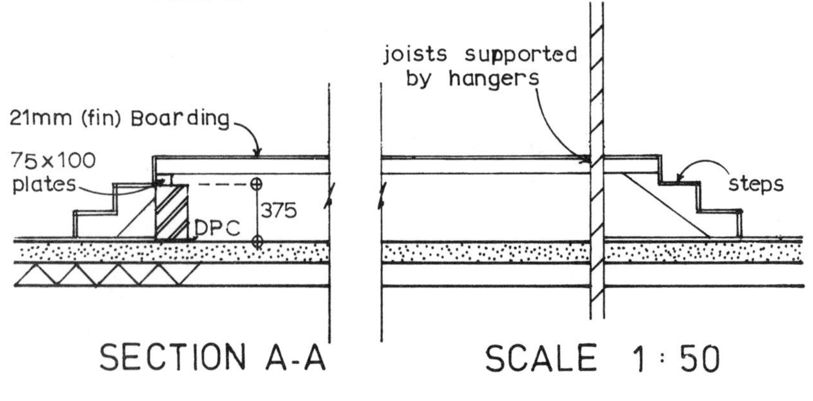

FIGURE 24

Suspended Ground Floor

FLOORS 1.

8.00 0.38		Brick walls, facework one side, one brick thickness, vertical, English bond in gauged mortar (1:1:6), including pointing with a flush joint as the work proceeds one side	F10.1.2.1
2/ 8.00 0.38		Brick walls, commons, half brick thickness, vertical, built honeycombed in gauged mortar (1:1:6)	F10.1.1.1 SMM does not mention honeycomb walls but obviously this must be stated and could be equated to type of bond.

			FLOORS 2.	
	8.00 0.23 2/ 8.00 0.11		Damp proof courses, width not exceeding 225mm, horizontal, of Hyload pitch polymer, bedding in gauged mortar (1:1:6)	F30.2.1.3 The d.p.c. to the partition would be measured with the partition.
2/3/	1		Air bricks, square hole pattern, clay, size 225 × 150 in hollow wall comprising half brick facings, 150 block inner skin and 50 cavity including sealing cavity with slates in gauged mortar (1:1:6)	F30.12.1 The formation of the opening is deemed to be included but the wall has to be described. If a lintel or arch is required this would be included in the description.

FLOORS 3.

			8000 Halving joint 150 8150	G20.8.0.1 The maximum length of timber available at normal cost is about 5.1 m. If it is accepted that lengths can be butt jointed then the net length is measured. If halving joints are required 150 should be added for each joint.
3/	8.15		Plates 100 x 75 treated sawn softwood	Bedding plates is deemed included.
			[Joists. 8000 Clearance 25 ½ joist 19 - 2/44 88 ÷350) 7912 = 23 +1 Number of jsts. = 24	The division is taken to the next whole number. Add 1 to the number of spaces to find the number of joists.

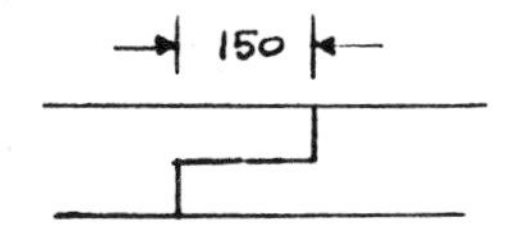

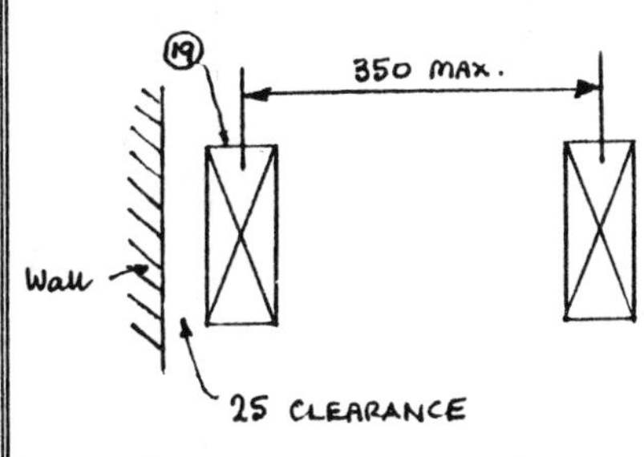

FLOORS 4.

Timesing	Dimensions	Squaring	Description	Notes
			5250 18 : 215 5465 Lapping at plate 100 5565	As the joist length is over 5.1 m an allowance of 100 has been made for lapping over a plate. JOIST · 100 · JOIST · PLATE
24/	5.57		Floor members 38 x 100 treated Sawn Softwood	G20.6.0.1
			[Hangers.	
24/	1		3mm Welded galvanised mild steel regular masonry hanger for 38 x 100 joist	G20.21.1 Fixing is deemed included.
			5465 ½ Riser 10 5475	NOSING · TOP RISER · 19
	8.00 5.48		Timber board flooring width exceeding 300, 21 (finished) wrot Softwood tongued & grooved	K20.2.1.1
	8.00		Nosing 38 x 25 wrot Softwood, twice rounded, tongued to edge softwood floor including groove	P20.2.1 Method of jointing stated Ends etc. deemed included.

EXAMPLE 6

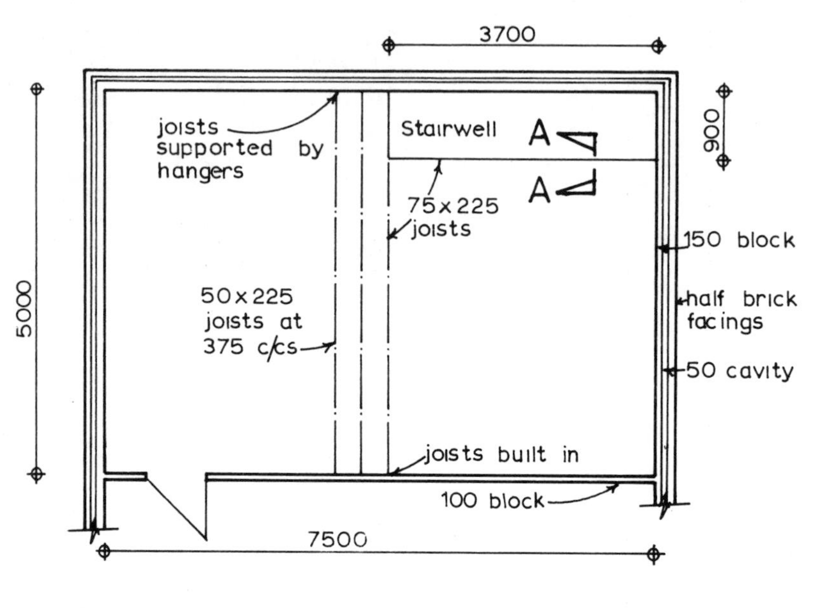

PLAN SCALE 1:100

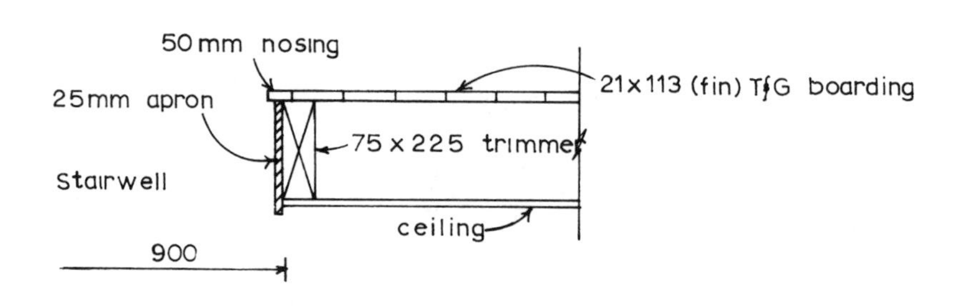

SECTION A-A SCALE 1:20

TIMBER UPPER FLOOR

FIGURE 25

Timber Upper Floor

FLOORS 1.

[Construction.

$\frac{1}{2}$/75 = 3700
37½
3737½

Clearance 25
Joist $\frac{1}{2}$/50 = 25 – 50

375) 3687½
= 10 + 1

11 (incl. trimming)

7500
– 3700
3800

As above 50
$\frac{1}{2}$/75 = 37½ – 87½

375) 3712½

10

Length 5000
100
5100

10/11	5.10	Floor members 50 x 225 treated Sawn softwood

Where a joist is in a fixed position, such as the trimming joist shown, the calculation for the number of joists should be made on either side of the fixed position.

G20.6.0.1

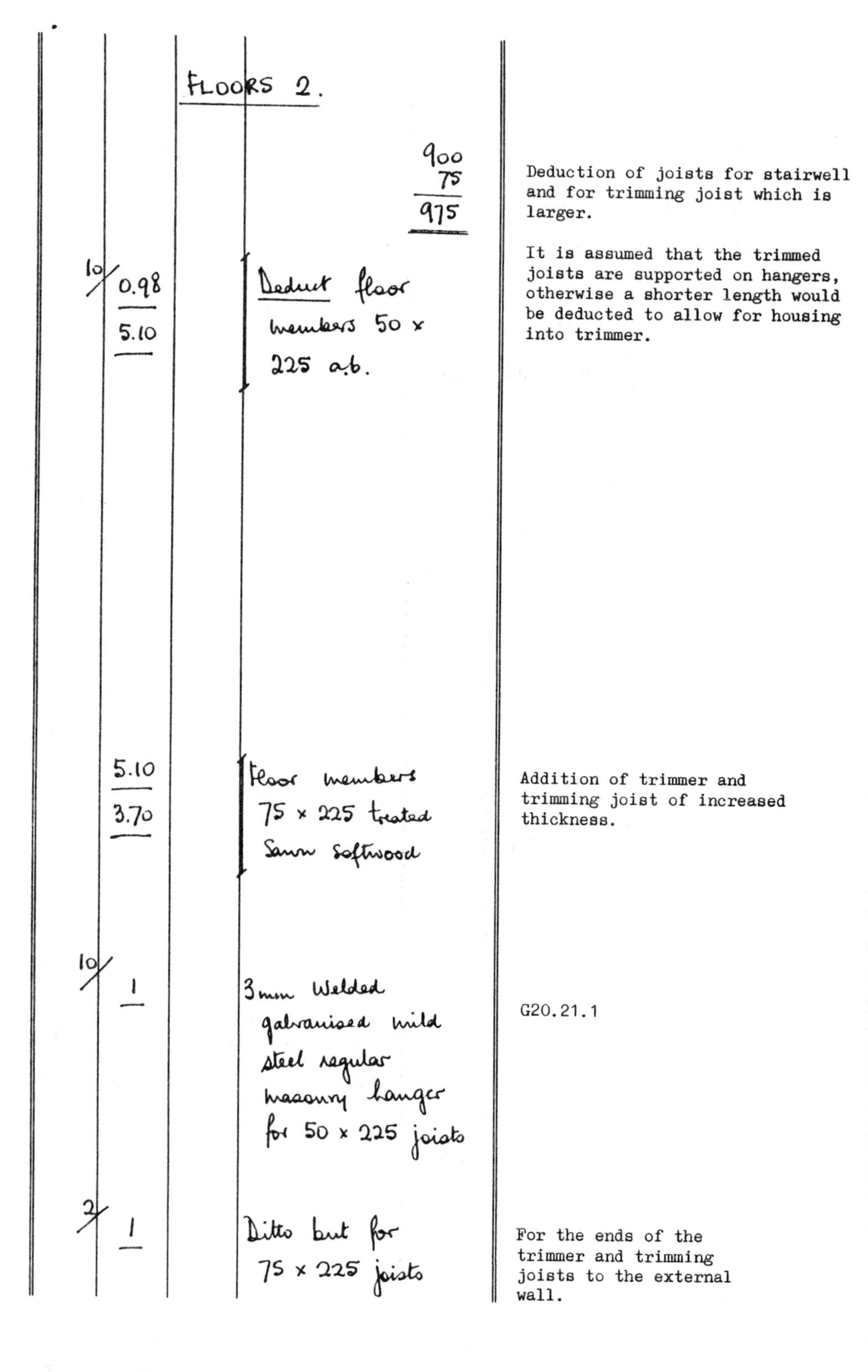

Deduction of joists for stairwell and for trimming joist which is larger.

It is assumed that the trimmed joists are supported on hangers, otherwise a shorter length would be deducted to allow for housing into trimmer.

Addition of trimmer and trimming joist of increased thickness.

G20.21.1

For the ends of the trimmer and trimming joists to the external wall.

FLOORS 3.

Timesing	Dimension	Squaring	Description	Notes
10/	1		3mm Welded galvanised mild steel regular timber to timber hanger for 50 x 225 joists	For the ends of the trimmed joists.
	1		Ditto but for 75 x 225 joists	
	7.50		Joist strutting, herringbone, 38 x 19 treated sawn softwood to 225 deep joists	G20.10.1
2/2/	1		Galvanised mild steel regular joist restraint strap 30 x 5 x 1000 girth, once bent & three times countersunk drilled	G20.20.1 Restraint straps should be taken every 2 m where joists run parallel to an external wall. Fixing is deemed to be included.

	Dimensions		Description	Notes
			FLOORS 4.	
	7.50 5.00		Timber board flooring width exceeding 300, 21 x 113 (finished) wrot softwood tongued and grooved	K20.2.1.1
			3700 + 10 = 3710 900 + 10 = 910	
	3.71 0.91		Deduct	
	3.71 0.91		Nosing 50 x 25 wrot softwood twice rounded, tongued to edge of softwood floor including groove	P20.2.1 Method of jointing stated. Ends etc. deemed included.

50
NOSING
SAY 10
APRON LINING
JOIST
OPENING SIZE

FLOORS 5.

			225
		Cg.	9
		Clearance.	10
			244

3.70 0.90	Apron lining 250 x 25 wrot softwood	P20.1.1 Ends etc. deemed included.
	&	
	K.p.s. & two undercoats & one finishing coat general surfaces woodwork isolated surfaces girth not exceeding 300 in staircase areas	M60 Work in staircase areas to be given separately (M1)

EXAMPLE 7

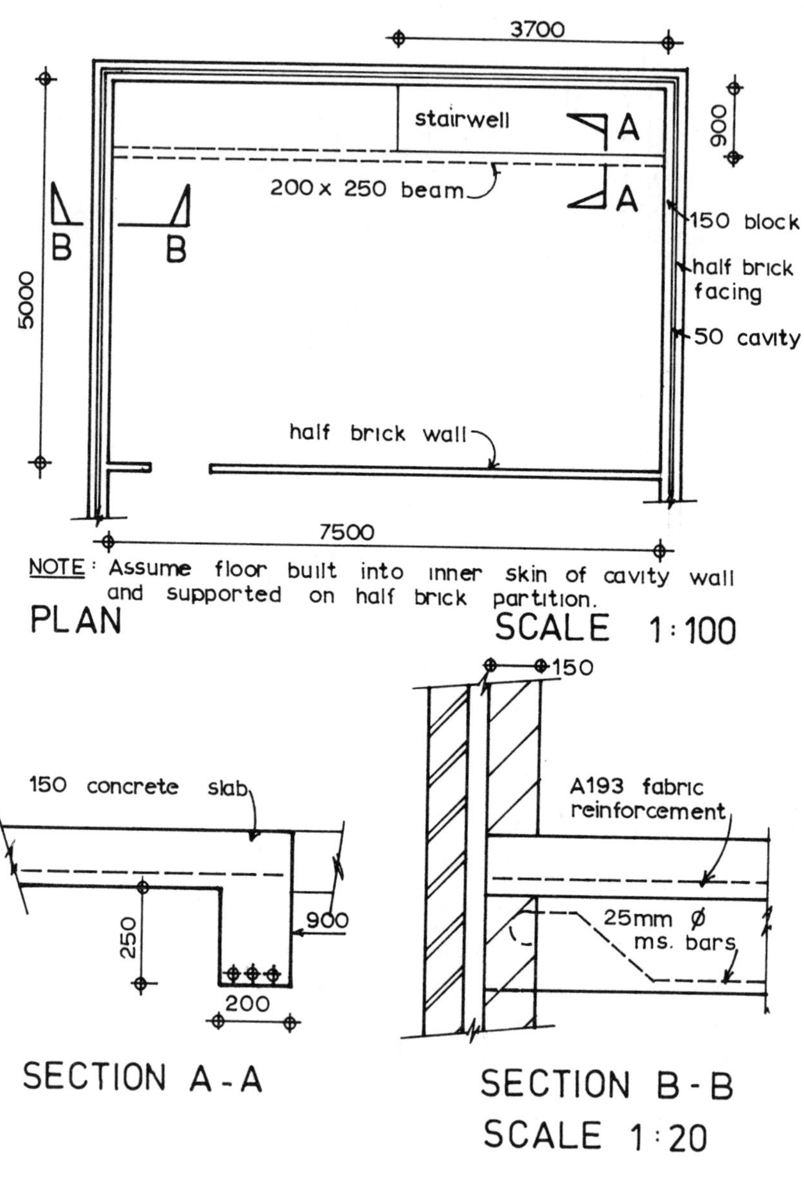

CONCRETE UPPER FLOOR

FIGURE 26

Concrete Upper Floor

FLOORS 1.

		7500	5000
	2/150 :	300	150
			102½
		7800	5252½

Dimensions	Description	Notes
7.80 5.25 0.15	In-situ concrete (25 N/mm² 20mm aggregate) slabs, thickness not exceeding 150, reinforced	E10.5.1.0.1
7.80 0.20 0.25		Volume of beam included in slab.

3700	900
150	150
3850	1050

Stairwell.

Dimensions	Description
3.85 1.05 0.15	Deduct

FLOORS 2.

	7.50 5.00		Formwork for in-situ concrete to soffits of slabs, slab thickness not exceeding 200, horizontal, propping 1.50 - 3.00m	E20.8.1.1 Formwork is measured to the area requiring supporting and is therefore not required where the slab is built into the wall.
			[Stairwell.	
	3.70 0.90		Deduct	
	7.50 0.20		[Beam.	
			200 2/250 = 500 700	
	7.50 0.70		Ditto to beams attached to in-situ slabs, regular shaped, propping 1.50 - 3.00m (In Nr. 1)	E20.13.1.1 Formwork to edge of suspended slab associated with beam is added. (M12)
	3.70 0.15		[Edge of slab	Formwork to ends of beams is deemed included. (C3)

FLOORS 3.

2/	7.80		Formwork for in-situ concrete to edges of suspended slabs, plain vertical, height not exceeding 250	E20.3.1.2 Formwork to edge of slab for entire perimeter.
2/	5.25			
	3.85		Deduct	Deduction where beam is below the edge of the slab at stairwell. (Previously measured with beam formwork) Note. Short edge of stairwell = edge of slab taken above.

FLOORS 4.

[Reinforcement.

		7800	5252½
Cover - 2/25 :		50	50
		7750	5202½

Dimensions	Description	Reference
7.75 5.20	Reinforcement for in-situ concrete, welded fabric, Reference A193, weighing 3.02 kg/m² with 200 laps	E30.4.1
	[Stairwell	
3.85 1.05	Deduct	

FLOORS 5.

[Beam reinf.

	7800	
- Cover 2/25	50	Allowance for hooks = 9 x diameter (rounded to next 10.)
	7750	
Hooks 2/230 =	460	
	8210	
Extra for Cranks say 2/80	160	
	8370	

3/	8.37	Reinforcement for in-situ concrete, bar, 25 bent as BS 4449 hot rolled plain steel	E30.1.1.2
	7.50 5.00	Trowelling surface of unset concrete to receive paving	E41.3
		[Stairwell.	
	3.70 0.90	Deduct	

10 Roofs

The measurement of roofs is sub–divided conveniently into two main sections of coverings and construction. As with floors, if the coverings are measured first it should be easier to understand the construction. When a building has different kinds of roof, e.g. tiled roofs and felt flat roofs, each should be dealt with separately. If there are several roofs of similar type it may be more convenient to group these together. The rainwater system, unless designed with the plumbing installation, is measured with the roofs section, and may either be taken at the end of each type of roof or at the end of the entire roof measurement.

If no roof plan is provided this may be superimposed on the plan of the upper floor. Special care must be taken to ensure that all hips and valleys are shown, one or the other being necessary at each change of direction in a pitched roof. In the case of a flat roof it is necessary to show the direction of fall, gutters and outlets.

TILE OR SLATE ROOF COVERINGS

The first measurement in this type of roof should be the area of coverings, attens and underlay being included in the description. If the area of one
pe is entered in the dimensions care must be taken to times this by two
low for both slopes. Following the coverings one should proceed
the edges of the roof, measuring abutments, eaves, verges, ridges,
valleys. The adjustments for chimneys and dormers and any
metal gutters, soakers and flashings could conclude the
asurement or be taken as a separate section at the end after
n. The calculation of the lengths of roof slopes, hips and
dealt with in Chapter 4.
specified, customary girths for leadwork to pitched

• cover flashings	150 mm
• stepped flashings (over soakers)	200 mm
• ditto (without soakers)	300–350 mm
• aprons	300 mm
• soakers 200 mm wide	length = gauge + lap + 25 mm (number = length of slope ÷ gauge)

PITCHED ROOF CONSTRUCTION

The notes on timber sizes and the measurement of timber floor construction in the previous chapter should be understood before reading this section. With traditional timber roof construction a logical sequence similar to the following should be adopted.

	SMM Reference
Plates	G20.8
Rafters	
Ridge	
Hips and valleys	G20.9
Ceiling joists and collars	
Purlins, ceiling beams, hangers and struts	
Sprockets	G20.17
Adjustments for openings, dormer construction etc. (as above)	
Ties to walls	G20.20
Walking boards	K20.2
Insulation	P10.2
Ventilation	H60.10

The number of rafters in a gabled roof should be calculated by taking the total length of the roof between the centre lines of the end rafters and dividing this length by the rafter spacing. The result should be rounded to the next whole number and one added to convert the number of spaces to the number of rafters. The number of rafters will have to be timesed by two for both slopes. If an intermediate rafter such as a trimmer is in a fixed position then the calculation should be made on either side of the fixed rafter. If the end of the roof is hipped then a similar calculation can be made taking the dimension from the foot of the jack rafters at the hipped end. One extra rafter should be taken at each hipped end opposite the ridge as shown in the sketch.

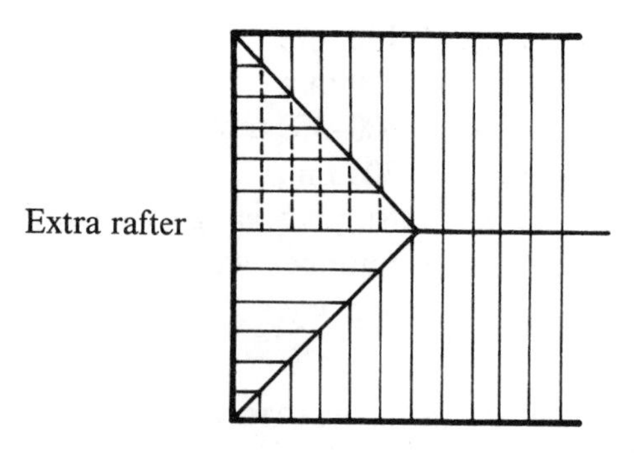

Fig. 27

It will be seen from the sketch that the jack rafters at the hipped end are equal to the lengths of the common rafters (shown dotted) which would have been there if the roof were gabled.

The number of ceiling joists may be calculated in a similar way, but the length to be divided is taken between the inside face of the external walls less the clearance as shown for timber floor construction.

All structural timber in pitched roofs is measured linear and described as in pitched roofs giving the size in the description. Plates and bearers are kept separately and described as plates. Cleats and sprockets are enumerated, their length and size being given in the description. Metal connectors, straps, hangers, shoes, nail plates, bolts and metal bracing are enumerated.

TRUSSED RAFTERS

Trussed rafters are enumerated and described and although there is no specific requirement in the SMM for a dimensioned diagram to be given, this is often the clearest way to describe a truss. Bracing and binders to the construction are taken as linear items. When calculating the number of trussed rafters consideration should be given to the position of water storage cisterns as their support may require additional trusses, platforms and bracing. This will probably involve discussion with the surveyor measuring the plumbing installation and agreement should be reached on where additional items will be measured.

EAVES AND VERGE FINISH

Verge, eaves, fascia and barge boards are measured linear giving their size in the description. Those exceeding 300 mm wide may be measured superficial. Painting the boards should be taken at the same time as a superficial item unless the member is isolated and less than 300 mm wide when it is measured linear.

RAINWATER INSTALLATION

Rainwater gutters are measured linear over all fittings, these being enumerated and described as extra over the gutter. Rainwater pipes are measured in a similar manner and, in addition, stating if they are fixed in ducts, chases, screeds or concrete. The description of gutters and rainwater pipes should include their size, method of jointing, type of fixing and if to special backgrounds. Rainwater heads are enumerated and gratings may be included in the description.

A careful check should be made to ensure that all roof slopes are drained, that there are sufficient rainwater pipes and that the drainage plan shows provision for each pipe. If the sizes of gutters and pipes are not shown then these have to be calculated, but it should be remembered that a variety of sizes, although theoretically correct, does not always lead to economy.

FLAT ROOFS

The same main subdivision of coverings and construction can be made as before. The examples show the measurement of asphalt flat roofs. The measurement of metal covered roofs should not prove too difficult provided that their construction is fully understood.

Metal roofs are measured their net visible area plus additions for drips, welts, rolls, seams, laps and upstands as detailed in the SMM. No deductions are made for voids not exceeding 1 m^2 within the area of the roof. When the opening has not been deducted no further work around the opening has to be measured except in the case of holes. The pitch of the roof has to be stated in the description.

Flashings are measured linear stating the girth and dressing into grooves, ends, angles and intersections are deemed included. Gutters are measured in a similar manner and labours are deemed to be included.

Cesspools are enumerated stating their size and again labours are deemed included.

Asphalt roofing is measured the area in contact with the base and again no deductions are made for voids within the area not exceeding 1 m^2. The pitch has to be stated in the description. The width of work has to be classified as not exceeding 150 mm, 150 to 225 mm, 225 to 300 mm, and exceeding 300 mm. Skirtings, fascias, aprons, channel and gutter linings, and coverings to kerbs are measured linear and classified in the same widths or girths, the exact girth being given if over 300 mm. Most labours to these items are deemed included. Internal angle fillets, fair edges, rounded edges, drips and arrises (angles) on main areas are, however, measured as linear items.

Felt roof coverings are measured superficial stating the pitch. Skirtings, gutters, coverings to kerbs, etc. are measured linear when not exceeding 2 m girth and the girth is given in 200 mm stages. Linings to cesspools, outlets and dishing to gullies are enumerated.

Timber flat roof construction is measured in a similar manner to that of floors. The fall in the roof may be created by the use of firrings which are measured linear giving their thickness and mean depth. Gutter boards and sides are measured superficial classifying the width as not exceeding or exceeding 300 mm. Timber drips and rolls, if required for joints in metal coverings, are measured linear stating their size. Fascias, soffits and rainwater goods are taken in a similar manner to those for pitched roofs.

Concrete roofs are measured in a similar manner to floors the fall usually being created in the screed. If the slab is sloping then this is classified as either not exceeding or exceeding 15 degrees. If the slab exceeds 15 degrees formwork has to be measured to the upper surface. Concrete upstands are measured as cubic items, formwork to the sides, if under 1 m, is measured linear and classified as not exceeding 250 mm, 250 to 500 mm and 500 mm to 1 m. Trowelling the surface of the concrete is measured superficial.

Taking–off List	**SMM Reference**
Pitched roof coverings:	
Tiling or slating (battens & underlay)	H60.1
Abutments	H60.3
Eaves	H60.4
Verges	H60.5
Ridge	H60.6
Hips	H60.7
Valleys	H60.9
Eaves and verge finish and rainwater disposal:	
Fascia	G20.15
Soffit	G20.16
Barge boards	G20.15
Decoration	M60.1
Gutters	R10.10
Fittings to ditto	R10.11.2
Rainwater pipes	R10.1
Fittings to ditto	R10.2.3/4
Connection to drain	R10.2.2
Timber pitched roof construction:	
See Page 153	
Substitute if necessary trussed rafters,	G20.2
binders and bracing	G20.9.2
Projections through pitched roofs:	
Deduct coverings (openings over 1 m^2)	As before
Abutments	H60.3
Eaves (above opening)	H60.4
Gutter and lier boards	G20.14
Metal gutters and flashings	H71–75.19/10
Stepped flashings	H71–75.10
Soakers	H71–75.26
Apron flashings	H71–75.11
Adjust construction	As before

Taking-off List (cont.) **SMM Reference**

Flat roof coverings:

	METAL	ASPHALT	FELT
Screeds	N/A	M10.6.1	M10.6.1
Coverings	H70–75.1	J21.3	J41.2
Skirtings and upstands	H70–75.1	J21.5	J41.10
Flashings	H70–75.10	N/A	J41.11
Kerbs	H70–75.16	J21.11	J41.14
Edges	H70–75.23	J21.13/14	N/A
Fascias	H70–75.1	J21.6	N/A
Channels	N/A	J21.9	N/A
Gutters	H70–75.19	J21.8	J41.13
Cesspools	H70–75.20	J21.19	J41.15
Outlets	H70–75.22	included	J41.18

Timber flat roof construction:	
Wall plates	G20.8
Timber beams	G20.9.1
Joists	G20.9.1
Joist hangers and connectors	G20.21/24
Strutting	G20.10
Ties to walls	G20.20
Firrings	G20.13
Boarding	K20.4
Drips and rolls	G20.13
Gutters	G20.14
Insulation	P10.2
Vapour barrier	P10.1

Concrete flat roof construction:	
Slab	E10.5
Beams	E10.9
Upstands	E10.14
Formwork	E20.8/13
Reinforcement	E30.1/4
Insulation	P10.2
Vapour barrier	P10.1
Trowelling slab	E41.3

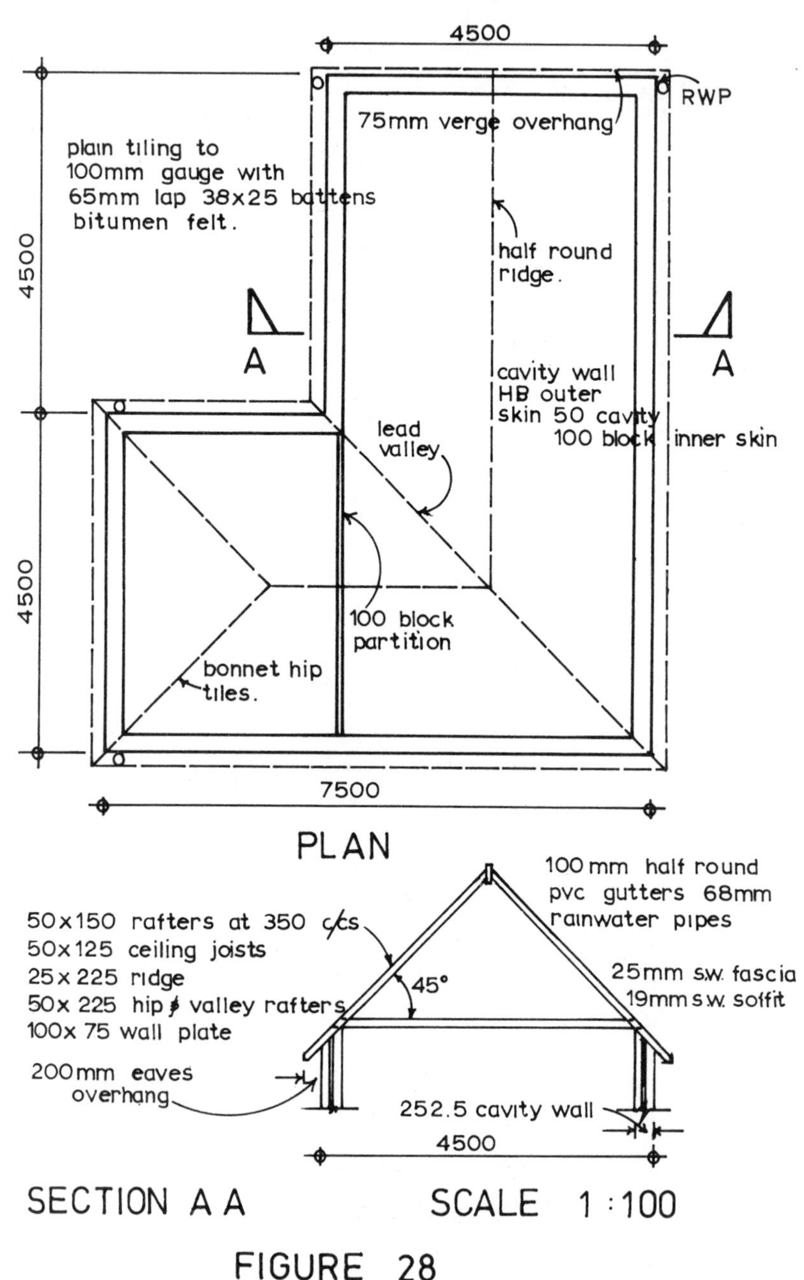
EXAMPLE 8
TRADITIONAL PITCHED ROOF
4500
RWP
75mm verge overhang
plain tiling to
100mm gauge with
65mm lap 38x25 battens
bitumen felt.
half round
ridge.
4500
A
A
cavity wall
HB outer
skin 50 cavity
100 block inner skin
lead
valley
4500
100 block
partition
bonnet hip
tiles.
7500
PLAN
100 mm half round
pvc gutters 68mm
rainwater pipes
50x150 rafters at 350 c/cs
50x125 ceiling joists
25 x 225 ridge
50x 225 hip & valley rafters
100x 75 wall plate
45°
25mm s.w. fascia
19mm s.w. soffit
200mm eaves
overhang
252.5 cavity wall
4500
SECTION A A
SCALE 1 : 100

FIGURE 28

Traditional Pitched Roof.

ROOFS 1.

Coverings.

		7500
Overhang.	200	
Into gutter.	50	
	2/250 =	500
		8000
		4500
		4500
		9000
Eaves a.b.	250	
Verge.	75	325
		9325

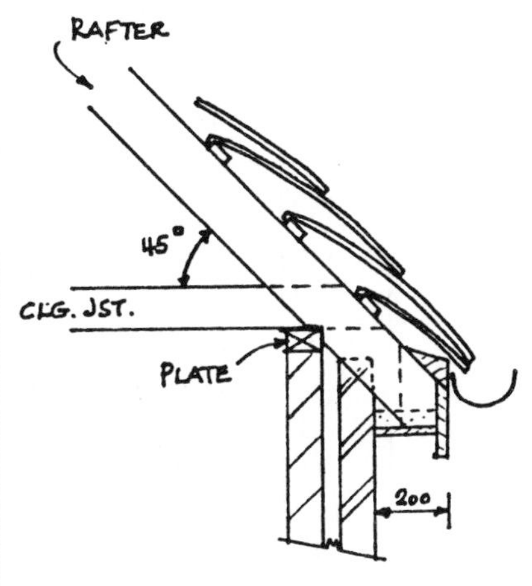

Waste calculation for overall plan area of roof.

Sec 45°/ 8.00
9.33

Roof coverings, 45° pitch, hand made sand faced plain clay tiles, 65 lap, each tile in every 4th course fixed with two aluminium nails, 38 x 25 treated softwood battens galvanised nailed to 100 gauge, reinforced bitumen felt to BS 747 type 1F weighing 15kg/10m² with 75 horizontal & 150 end laps

H60.1.1

The area of the roof coverings equals the plan area multiplied by the secant of the slope.

ROOFS 2.

		4500 - Eaves a.b. 250 4250 Verge 75 4325 7500 - 4500 3000	
Sec 45°	4.33 3.00	Deduct Roof Coverings a.b.	
		2/8000 = 16000 2/9325 = 18650 34650 2/250 = 4500 500 5000 29650	
	29.65	Eaves, double tile course	H60.4
		[Tilting fillet. 29650 - 2/2/50 200 29450	Overhang into gutter deducted to obtain length of tilting fillet. 3 external angles - 1 internal = 2 for deduction.
	29.45	Individual support 75 × 50 (extreme) treated sawn softwood triangular	G20.13.0.1

Area to deduct

ROOFS 3.

		Description	Notes
		[Verge.	
Sec 45°/	5.00	Verge with plain tile undercloak	H60.5 Whilst C2 states that undercloaks are deemed to be included S5 asks for the method of forming to be stated.
		[Ridge.	
		8000 9325 17325 $-3/\frac{1}{3}/5000$ 7500 9825	
	9.83	Ridge 250 diameter, half round, red clayware	H60.6
3/	1	Saddle 450 x 450 Code 4 lead, handed to others for fixing & Fixing only ditto	H71.25.1 Lead saddle at angle of ridge and at intersection with two hips. H60.10.6.1.1

ROOFS 4.

$\frac{1}{2}/5000 = 2500$

$2500 \sec 45° = 3535$

= slope length

Hip length $= \sqrt{3535^2 + 2500^2}$

$= 4330$

3/	4.33	Hips, matching bonnet hip tiles	H60.7
		[Valley.	
	4.33	Gutter, 710 girth, sloping, Code 5 lead with 150 laps	H71.19.1.0.3 (C2a) Laps and ends deemed included but laps have to be specified. (S4)
		& [Gutter base.	
		Gutter board width not exceeding 300, 250 × 25 twice splayed treated wrot softwood	G20.14.3.1 Base to gutter.
2/	4.33	Valley edge	H60.9 As there are no valley tiles this item is taken so that the estimator is aware that a valley in the tiling exists which will require cutting normally included with boundary work.

ROOFS 5.

[Gutter sides.

Timesing	Dimension	Squaring	Description	Notes
2/	4.33		Gutter board width not exceeding 300, 200 x 25 splayed treated wrot softwood	G 20.14.3
			& [Tilting fillets.	
			Individual support 75 x 50 (extreme) treated sawn softwood triangular	G 20.13.0.1

[Eaves finish.

[Fascia.

4500		
4500		
2/9000 :	18000	Calculation for length of fascia.
2/7500 :	15000	
Brick face	33000	
Extl. Ls 2/2/200 :	800	One external angle compensates for one internal angle and therefore two external angles are added.
Verge 2/75 :	150	
	33950	
- Gable	4500	
	29450	

			ROOFS 6.	
	29.45		Fascia board, 25 × 175 wrot softwood, chamfered & grooved	G20.15.3.1
			[Soffit.	
			33 000 Extl Ls. 3/2/200 = 1 200 Verge 2/75 = 150 34 350 – Gable 4500 29850	Calculation for length of soffit. As one edge is rebated and the angles are mitred, the full length of the timber required is measured. e.g.
	29.85		Eaves soffit board, 19 × 200 wrot softwood rebated	G20.16.3.1
			33 000 – 4500 350) 28500 = 82 + 1 = 83	Calculation for number of supports for soffit. Length of brickwork divided by rafter spacing.
			175 200 375	
83/	0.38		Individual support, 38 × 50 treated sawn softwood	G20.13.0.1

Chamfer

Groove

FACE OF WALL

SOFFIT BOARD

BRACKET

RAFTER

			ROOFS 7.	
2/	1		Spandril boxed end to eaves 25 mm wrot softwood size 200 x 270 overall	Filling to end of fascia and soffit at gable.
	29.45		Prime only general surfaces wood isolated surfaces, girth not exceeding 300, application on site prior to fixing	M60.1.0.2.4 Priming to back of fascia and soffit prior to fixing.
	29.85			
			[Spandril ends.	
2/	1		Ditto isolated areas not exceeding 0.5 m^2, application on site prior to fixing	
			Ends 2/200 = 29450 + 400 = 29850	
	29.85		Knot, prime, stop, two undercoats, one finishing coat gloss paint, general surfaces wood, girth exceeding 300, externally	M60.1.0.1 Painting to fascia and soffit.
	0.40			

ROOFS 8.

Timesing	Dimension	Squaring	Description	Notes
			[Gutter.	
			Fascia 29450 3/2/100 = 600 30050	Extreme length of gutter measured over fittings.
	30.05		Rainwater gutters, straight, 100 half round, u.p.v.c. to BS 4576 with combined fascia brackets & clips screwed to softwood	R10.10.1.1
2/	1		Extra over ditto for stop end	R10.11.2.1
4/	1		Ditto angles	
4/	1		Ditto outlets	
			[R.W.P.s Length assumed.	
4/	6.00		Rainwater pipes, straight, 68 Ø u.p.v.c. to BS 4576 with push fit socket joints & ear piece brackets at 2m centres plugged to masonry	R10.1.1.1 Length of pipe measured over fittings.

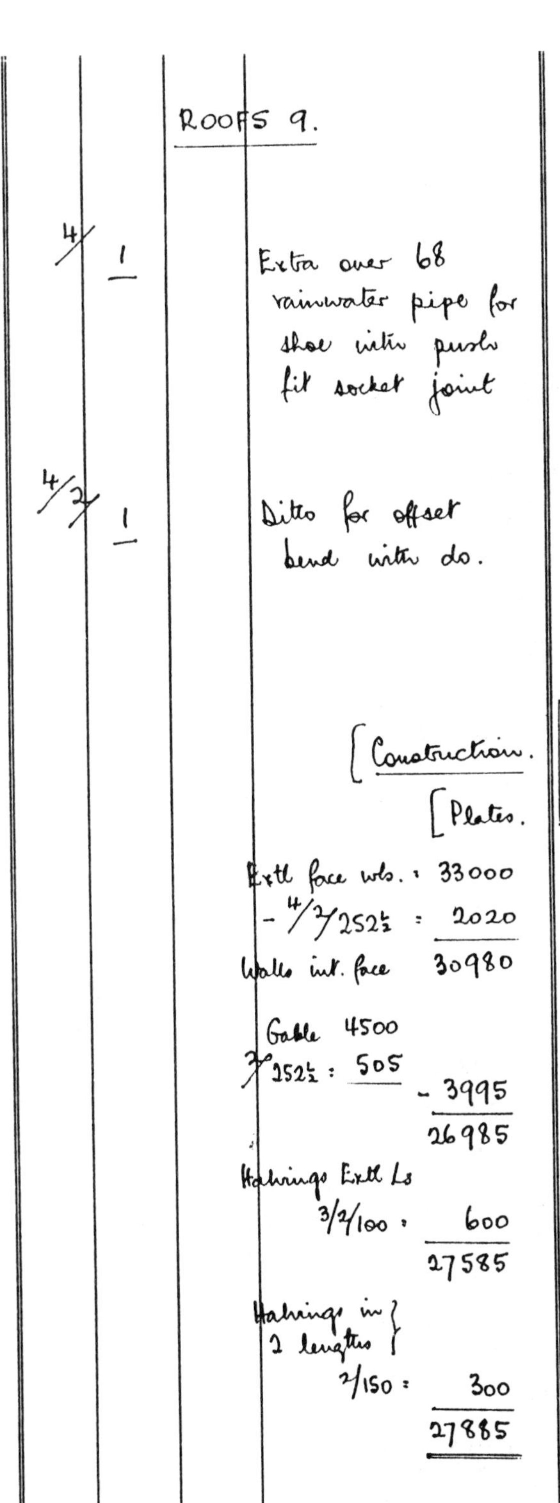

ROOFS 9.

4/ 1 Extra over 68 rainwater pipe for shoe with push fit socket joint

4/2/ 1 Ditto for offset bend with do.

[Construction.

[Plates.

Extl face wls. : 33000
- 4/2/252½ : 2020
Walls int. face 30980

Gable 4500
2/252½ : 505
- 3995
26985

Halvings Extl Ls
3/2/100 : 600
27585

Halvings in 2 lengths
2/150 : 300
27885

R10.2.4.6

Fittings to pipes over 65 ∅ have to be described.

Measure connection to drain here if appropriate.

R10.2.4.5

Bends to form swan-neck.

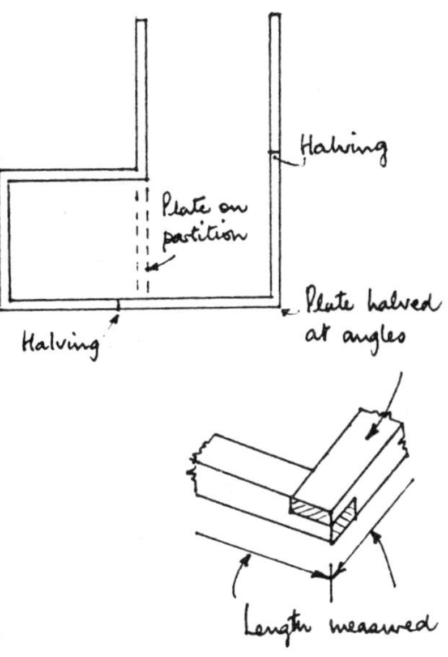

As two lengths are over 5.1 m, halving joints are allowed for in the lengths to avoid butt jointing.

ROOFS 10.

[Plates. (Contd.)

[Partition.

		4500
	− 2/252½ =	505
		3995
Halvings at ends	2/100 =	200
		4195

27.89		Plates 100 x 75 treated sawn Softwood	G20.8.0.1
4.20			

[Rafters

[Number.

		9000
Wall	252½	
Clearance	25	
½ Rafter	25	302½
		8697½
Overhang		200
		8897½
	÷ 350	= 26

	7500
−	4500
	3000
÷ 350	= 9

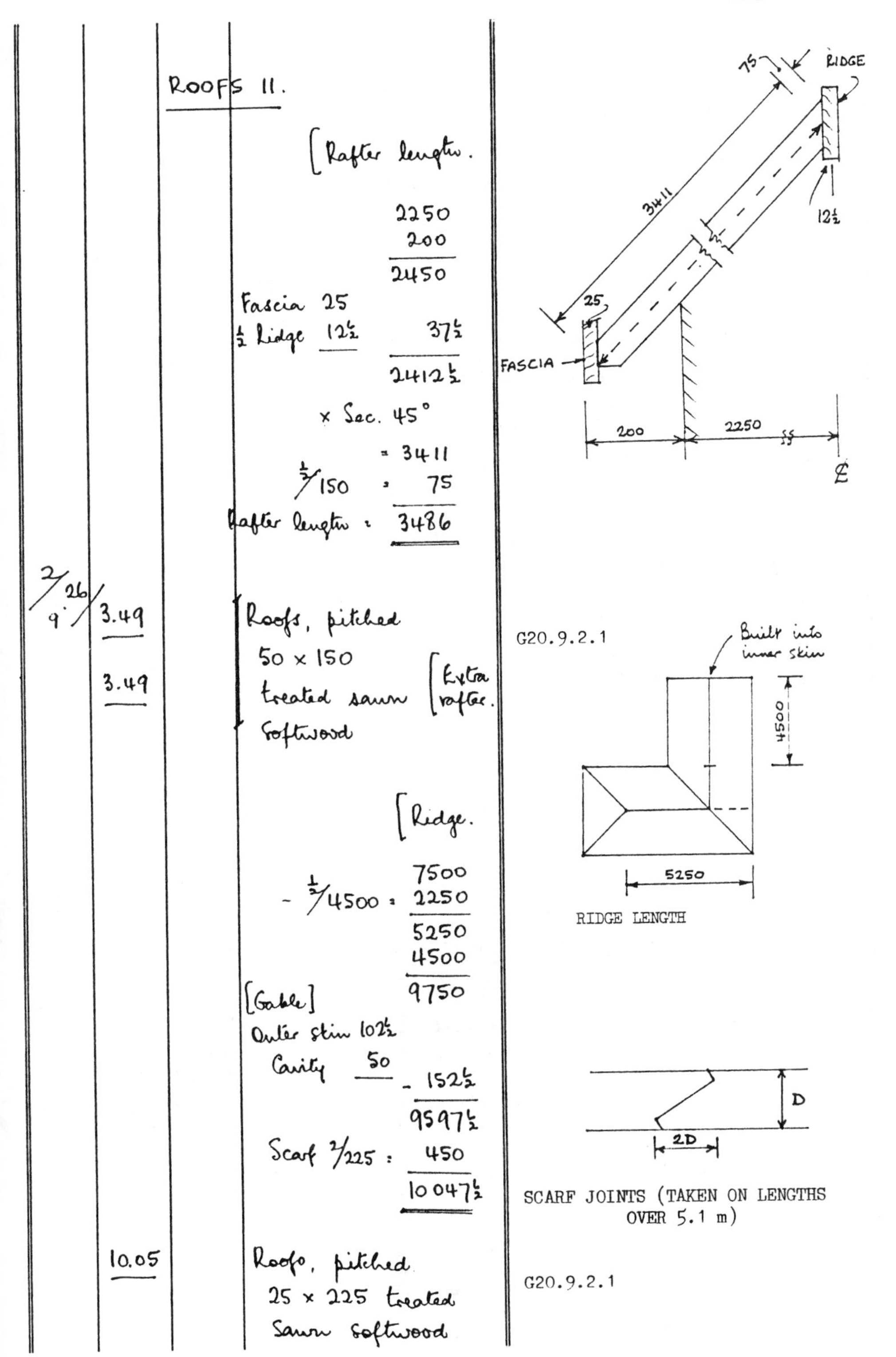
ROOFS 11.
[Rafter length.
2250
200
2450
Fascia 25
½ Ridge 12½ 37½
2412½
× Sec. 45°
= 3411
½/150 = 75
Rafter length = 3486
2/26/9 3.49
3.49
Roofs, pitched
50 × 150
treated sawn
softwood
[Extra
rafter.
G20.9.2.1
[Ridge.
7500
- ½/4500 = 2250
5250
4500
9750
[Gable]
Outer skin 102½
Cavity 50 - 152½
9597½
Scarf 2/225 = 450
10 047½
10.05
Roofs, pitched
25 × 225 treated
sawn softwood
G20.9.2.1
75
RIDGE
3411
12½
25
FASCIA
200
2250
℄
Built into
inner skin
4500
5250
RIDGE LENGTH
D
2D
SCARF JOINTS (TAKEN ON LENGTHS
OVER 5.1 m)

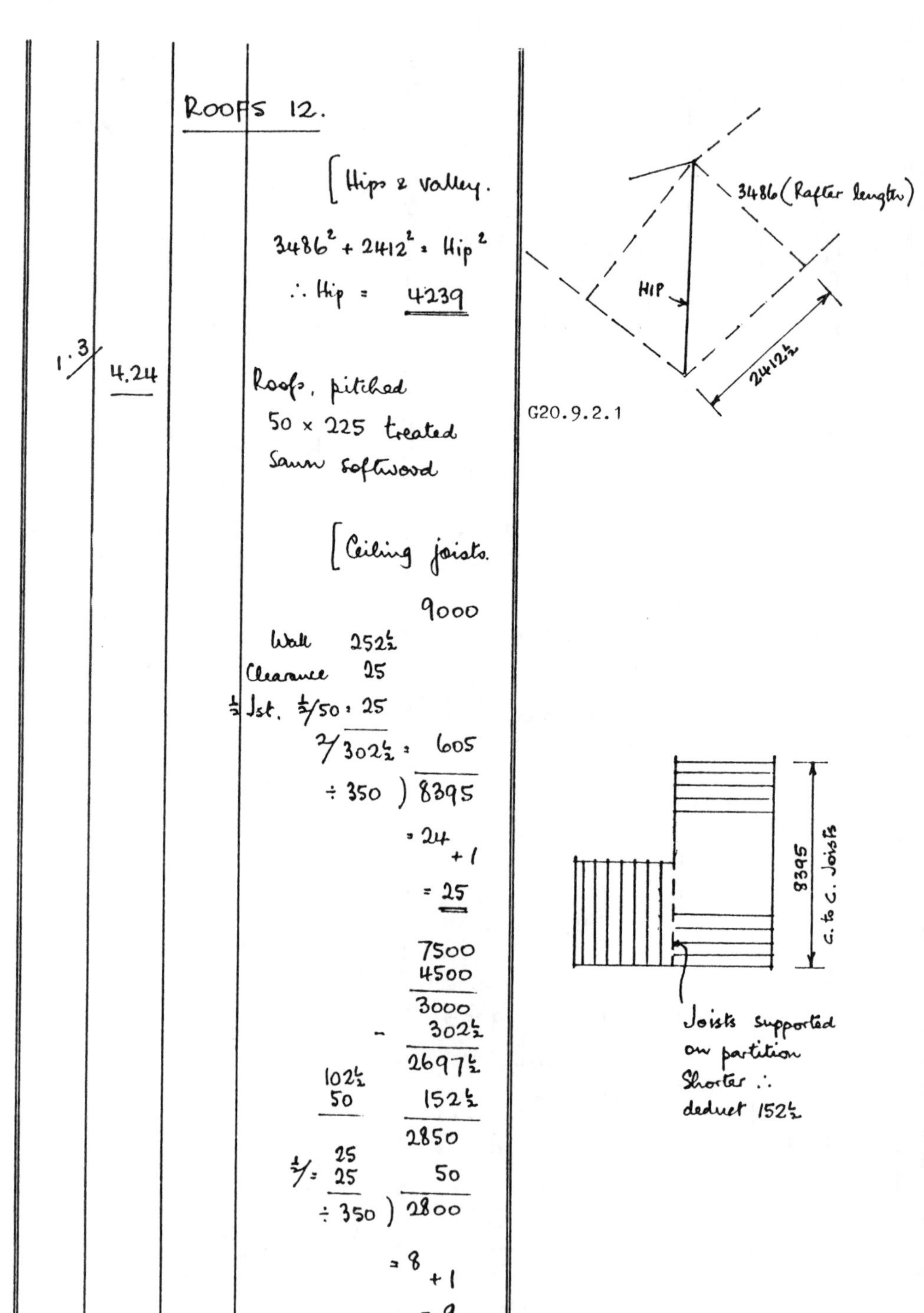

ROOFS 12.

[Hips & valley.

$3486^2 + 2412^2 = Hip^2$

∴ Hip = 4239

1.3/ | 4.24 | Roofs, pitched 50 × 225 treated sawn softwood | G20.9.2.1

[Ceiling joists.

9000

Wall 252½
Clearance 25
½ Jst. ½/50 = 25
2/302½ = 605

÷ 350) 8395
= 24
+ 1
= 25

7500
4500
3000
− 302½
2697½
102½
50
152½
2850
½/ = 25
25
50
÷ 350) 2800
= 8
+ 1
= 9

ROOFS 13.

9.25/	4.50		Roofs, pitched 50 x 125 treated Sawn softwood	G20.9.2.1
			4500 - 2/302½ 605 ÷ 350) 3895 = 11 + 1 = 12	NOTE: Insulation, eaves, ventilation, and wall anchors to take if required.
12/	0.15		Deduct [Joists on ptn.	

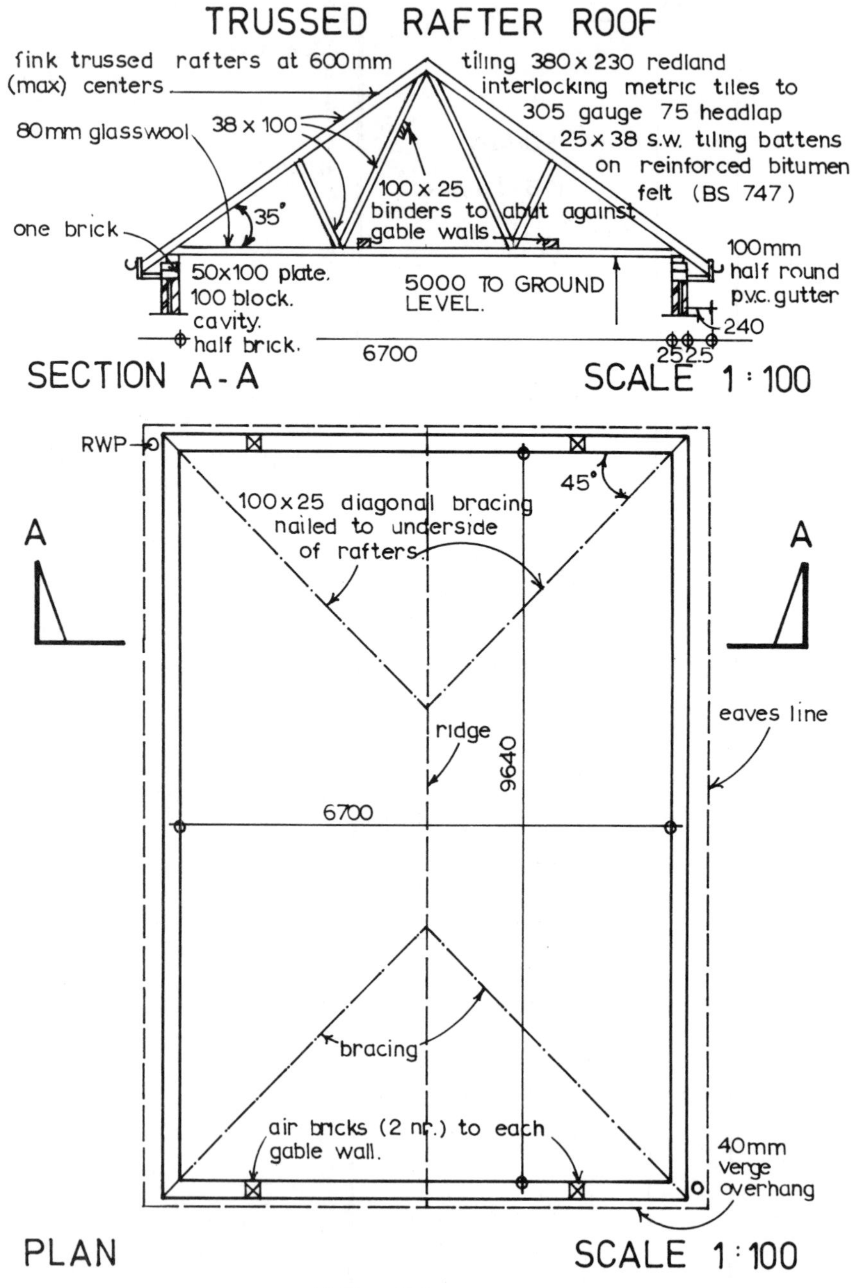
EXAMPLE 9
TRUSSED RAFTER ROOF
fink trussed rafters at 600mm (max) centers
tiling 380 x 230 redland interlocking metric tiles to 305 gauge 75 headlap
25 x 38 s.w. tiling battens on reinforced bitumen felt (BS 747)
80mm glasswool
38 x 100
100 x 25 binders to abut against gable walls
35°
one brick
50x100 plate.
100 block.
cavity.
half brick.
5000 TO GROUND LEVEL.
100mm half round p.v.c. gutter
240
6700
252.5
SECTION A-A
SCALE 1:100
RWP
100x25 diagonal bracing nailed to underside of rafters
45°
A
A
ridge
9640
eaves line
6700
bracing
air bricks (2 nr.) to each gable wall.
40mm verge overhang
PLAN
SCALE 1:100

FIGURE 29

Trussed Rafter Roof

ROOFS 1.

[Coverings.

		9640
Walls	2/252½ =	505
		10145
Verges	2/40 =	80
		10225

		6700
Wall	252½	
Eaves	240	
Into gutter	50	
	2/542½ =	1085
	½/	7785

= 3893 × Sec. 35°

Slope length = 4753

The slope length of the roof covering equals half the span multiplied by the secant of the pitch angle.

Timesing	Dimensions	Description
2/	10.23 4.75	Roof coverings 35° pitch, 380 × 230 Redland Interlocking metric tiles, 75 headlap, each tile in every alternate course nailed with one 48 × 11 gauge aluminium alloy nail, 38 × 25 treated sawn softwood battens galvanised nailed, reinforced bitumen felt to (Contd.)

H60.1.1

		ROOFS 2.		
			(Contd.) BS 747 type 1F weighing 15kg/10m² with 75 horizontal and 150 end laps	
2/	10.23		Eaves to tiling	H60.4
			& [Tilting fillet.	
			Individual support 75 × 50 extreme, treated sawn softwood, triangular	G20.13.0.1
			[Verge.	
2/2/	4.75		Verge, special tile including plain tile undercloak	H60.5
			[Ridge.	
	10.23		Ridge, 250 diameter half round to match general tiling	H60.6

ROOFS 3.

2/	10.15		Fascia board, width not exceeding 300, 25 x 150 wrot softwood, chamfered and grooved	G20.15.3.1
			&	
			25 − 13 = 12; 240 − 12 = 228	
			Eaves soffit board, width not exceeding 300, 25 x 228 wrot softwood, rebated	G20.16.3.1
			&	
			Individual support 38 x 50 treated sawn softwood	G20.13.0.1
2/2/	1		Spandril boxed end to eaves 25 wrot softwood size 240 x 300 overall	

38 x 50
FASCIA
WALL
SOFFIT

ROOFS 4.

Times	Dims		Description	Notes
2/2/	10.15		Prime only general surfaces of wood isolated surfaces, girth not exceeding 300, application on site prior to fixing	M60.1.0.2.4 Priming to back of fascia and soffit before fixing.
			[Spandril ends.	
2/2/	1		Ditto isolated areas n.e. 0.5m^2, application on site prior to fixing	M60.1.0.3.4 Ditto to spandril ends.
			Ends 2/300 10145 600 = 10745	
			250 150 25 = 425	
2/	10.75 0.43		Knot, prime, stop, two undercoats one finishing coat gloss paint, general surfaces of wood girth exceeding 300, externally	M60.1.0.1 Painting to fascia and soffit. This becomes a superficial item as the two combined girths are over 300.

		ROOFS 5.		
2/	10.23		Rainwater gutters, straight, 100 half round, UPVC to BS 4576 with combined fascia brackets & clips screwed to softwood	R10.10.1.1 The gutter is measured over fittings.
			[Stop ends.	
2/2/	1		Extra over for stop end	R10.11.2.1
			[Outlets.	
2/	1		Ditto for outlet	
			[RWPs.	
2/	5.20		Rainwater pipes, straight, 68 Ø UPVC with push fit socket joints & ear piece brackets at 2 m centres plugged to masonry	R10.1.1.1 The pipe is measured over fittings.

		ROOFS 6.		
2/2/	1		Extra over 68 rainwater pipe for offset bend with push fit socket joint	R10.2.4.5 Bends to form swan-neck. Fittings to pipes over 65 diameter have to be described.
2/	1		Ditto for connection to top inlet of clayware gully with cement & sand (1:3)	R10.2.2.1
			Construction.	
			[Plates.	
			9640 Build in 2/100: 200 9840 Halving 150 9990	As the total length is over 5.1 m, a halving joint is allowed for to avoid butt jointing.
2/	9.99		Plates 100 x 50 treated sawn softwood	G20.8.0.1

ROOFS 7.

[Trusses.

9640
½/38 = 19
25
-²/44 = 88
÷600) 9552
= 16 + 1

240
250
490
- 25
465

17/ 1

Trussed rafters in sawn softwood, FINK pattern, comprising 38 × 100 members, 6700 clear span, 465 eaves overhang, 35° pitch, jointed with 18g. galvanised steel plate connectors

G20.2.1

It may be appropriate to include a dimensioned diagram for this item.

ROOFS 8.

[Bracing etc.

		9640
	Lap.	100
		9740

[Binders.

1.2/	9.74	Roofs, pitched, 25 × 100 Sawn Softwood	G20.9.2.1

½/6700 = 3350
Plate 100
3450

3450 × Sec 35 = 4212

$3450^2 + 4212^2 = x^2$

$x = 5445$

[Braces.

2/2/	5.45	Roofs, pitched, 25 × 100 Sawn Softwood	
2/2/	1	Galvanised mild steel regular joist anchorage restraint strap 30 × 5 × 300 girth, twice bent and twice countersunk drilled	G20.20.1 Fixing of straps is deemed included.

ROOFS 9.

		2/ 6700 200 6900	
		9640 - Clg ties 17/38 : 646 8994	
	8.99 6.90	80 mm Glasswool insulation to BS 1785, between members at 600 centres, horizontal, lapped 100 mm at joints	P10.2.3.1
		[Air bks.	
2/2/	1	Air bricks, square hole pattern, red terra-cotta size 225 x 150 in hollow wall of half brick facings, 100 concrete block inner skin and 50 cavity including sealing cavity with slates	F30.12.1

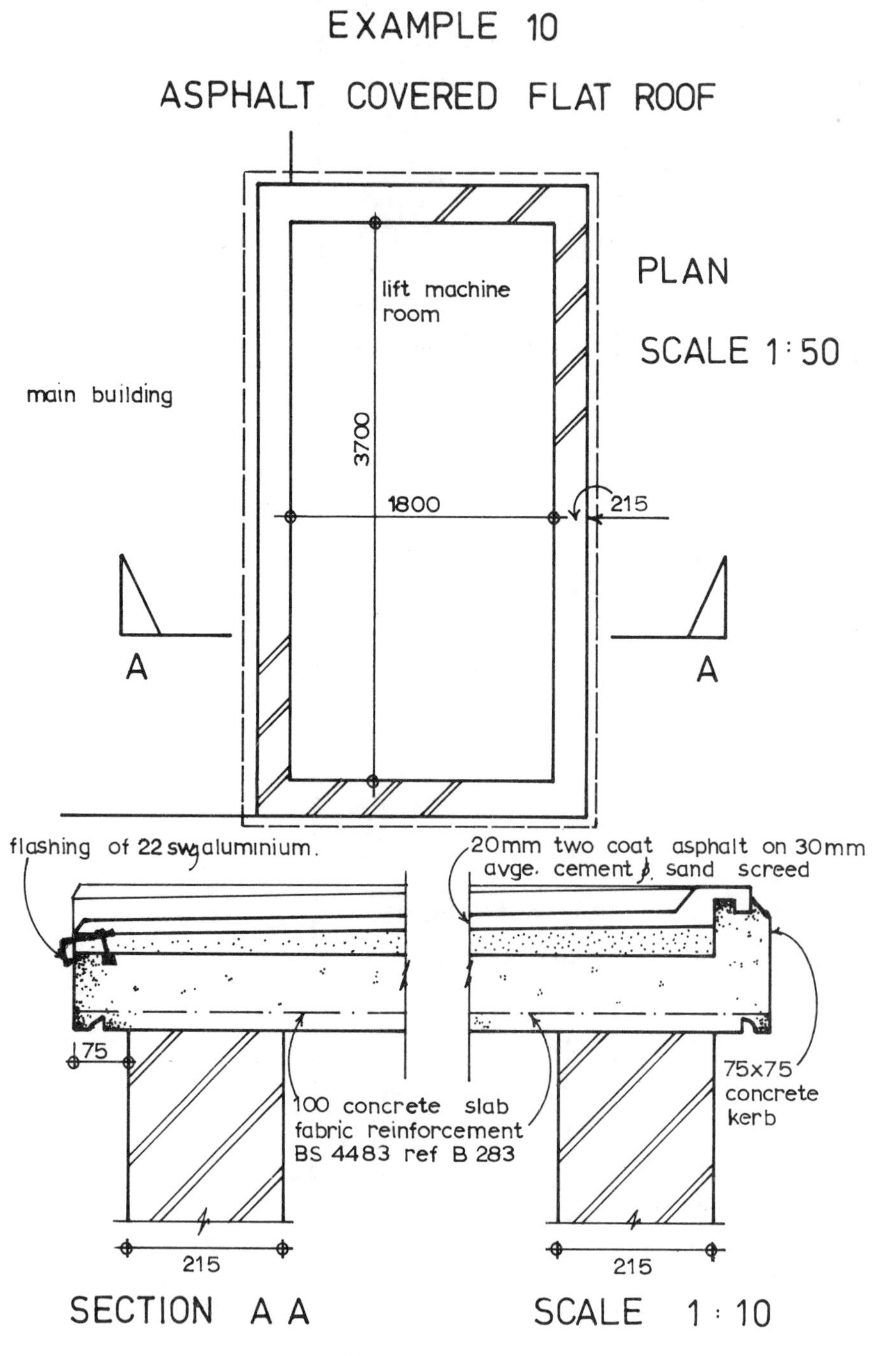
EXAMPLE 10
ASPHALT COVERED FLAT ROOF
lift machine room
PLAN
SCALE 1:50
main building
3700
1800
215
A
A
flashing of 22 swg aluminium.
20mm two coat asphalt on 30mm avge. cement & sand screed
75
100 concrete slab fabric reinforcement BS 4483 ref B 283
75x75 concrete kerb
215
215
SECTION A A
SCALE 1:10

FIGURE 30

Asphalt covered flat roof

ROOFS 1.

		1800	3700
	2/215 =	430	430
	2/75 =	150	150
		2380	4280

Dimensions	Description	Reference
2.38 4.28 0.10	In situ concrete (25 N/mm^2 - 20 mm aggregate) slab thickness not exceeding 150, reinforced	E10.5.1.0.1

	4280
2/2380 =	4760
	9040
- 2/75 =	150
	8890

Dimensions	Description	Reference
8.89 0.08 0.08	Ditto upstands	E10.14

ROOFS 2.

[Formwork

1.80 3.70 13.02 0.08	Formwork for in-situ conc. to soffits of slabs, slab thickness not exceeding 200, horizontal. Height to soffit 9.00 - 10.50 m	E20.8.1.1 Check the height of the formwork above a suitable base for propping. The height is required to be stated in 1.5 m stages. In this case the assumption is made that the slab in question is over a lift shaft.
	2380 4280 2/ 6660 = 13320 - 4/2/½/75 = 300 13 020	Calculation for girth of external soffit. The formwork for this has been added back to the last item but note the propping height may vary.

		ROOFS 3.		
	13.02 0.08		Extra over formwork to slabs for lining to produce fair finish	E20.20.1
	13.02		Ditto for recess to form quadrant throating 25 radius (In Nr 4)	E20.17.1.0.1

ROOFS 4.

Timesing	Dimensions	Squaring	Description	Notes
	13.32		Formwork for in-situ concrete to edges of suspended slabs, plain vertical, height not exceeding 250	E20.3.1.2 Formwork to the edge of the suspended slab including the upstand as the total depth does not exceed 250.
			100 75 175	
2/	2.38 0.18		Extra over ditto for lining to produce a fair finish	E20.20.1
	4.28 0.18			
	4.28 0.15			
			2/2380 = 4760 4280 9040 − 2/2/75 = 300 8740	
	8.74		Formwork for in-situ concrete to sides of upstands, plain vertical, height not exceeding 250	E20.4.1.2 Formwork to the inner face of the upstand.

ROOFS 5.

	$-2/2/\frac{1}{2}/75$: 9040 150 8890	
8.89	Formwork to form recess, plain rectangular 20 x 20 in top of concrete upstand (in Nr 3)	Similar to E20.17 Not taken as 'extra over' as no formwork measured to top of upstand.
9.04 0.30	Trowelling top of upstand to form weathering	E41.3 This could be measured as top formwork under E20.11 as the slope exceeds 15°. NOTE: The foregoing items relating to the formwork to the overhanging eaves and upstand are somewhat numerous and consideration should be given to measuring the item as a complex shape under E20.28 with a dimensioned description or diagram.

ROOFS 6.

[Reinforcement

		2380	4280
- Cover	2/35 =	70	70
		2310	4210

2.31 4.21	Reinforcement for in-situ concrete, welded fabric, Ref B 283 weighing 3.73 kg/m² with 200mm laps BS 4483 with plain hard drawn wire	E30.4.1

[Coverings

Mastic asphalt with limestone aggregate to BS 988 in two coats laid breaking joint by minimum 75mm and to total thickness of 20mm

This demonstrates the use of a heading to cover parts of a description which applies to several items.

ROOFS 7.

```
  2380          4280
-   75  -2/75:   150
  2305          4130
```

2.31 4.13	Asphalt roofing, level and to falls, width exceeding 300, to cement & sand base	J21.3.4
	&	
	Cement & sand (1:4) screed, roofs, level & to falls only, not exceeding 15° from horizontal, 30 (average) thickness, in one coat to concrete base	M10.6.1
	&	
	Trowelling surface of unset concrete to receive paving	E41.3

ROOFS 8.

8.89		Asphalt covering to kerb girth not exceeding 150 raking	J21.11.1 Coverings to kerbs are deemed to include arrises, internal angle fillets, turning nibs into grooves, end and angles so there is no need to describe these.
		3700 2/215 = 430 4130	
4.13		Fair rounded edge to asphalt & 22 SWG flashing of commercial purity aluminium, girth not exceeding 100, horizontal, fixed with galvanised nails to hardwood & Treated hardwood individual support 40 × 25 dovetail section, setting in surface of unset concrete	J21.13/14 (c2a) Work is deemed to include working to metal flashings H72.10.1.0.1 G20.13.0.1

11 Internal finishings

REVIEW OF WORK MEASURED

The measurement of roofs, covered in the last chapter, completes the work relating to the carcase of the building. The remaining work can broadly be described as the 'finishings' sections which are listed in Chapter 5. The taker–off at this stage, may find it advantageous to review the work completed so far, provided that it has been measured in the sequence suggested previously. A careful inspection of the drawings and specification and a look through the sub–headings of the appropriate sections of the SMM may reveal items which have been overlooked. Being satisfied that the carcase measurement is complete, one can then proceed to the next section of work which is internal finishings.

SCHEDULES

A schedule of internal finishings listing floor by floor and room by room; the ceiling, wall and floor finishes and decorations is an invaluable aid to the measurement of this section of the work. Details of cornices, skirtings, tiled dadoes, etc. can be added to the schedule as required. The schedule brings to light any missing information, enables rooms with similar finishings to be grouped together more easily and reduces the need to refer to the specification during the taking off. If the entries on the schedule are looped through as the items are measured one ensures that none is missed. At a later stage the schedule provides a useful quick reference indicating what has been taken without having to look through the dimensions. An example of a typical schedule is given in Appendix 2.

SUB–DIVISION

If the ceiling heights and construction vary throughout the building it may be advisable to keep the measurements on each floor separately. If, however the room layout and finishes repeat from floor to floor it is probably more expedient to use timesing. As this will quite possibly be the largest section of taking–off, care should be taken to signpost the dimensions to enable measurements to be traced easily at a later stage. The principal sub-divisions in the order recommended for measuring are:

a) floor finishings and screeds (unless taken with floor construction)
b) ceiling finishings (including attached beams with the same finish)
c) isolated beams
d) wall finishings (including attached columns with the same finish)
e) isolated columns
f) cornices
g) dadoes
h) skirtings

Note: It may be convenient to measure floors with ceilings if the areas are the same. Confusion may arise, however, if the specification changes are not consistent between the two and also if deductions have to be made from floors only for such items as hearths, stair openings, and fittings.

GENERALLY

The measurements for in situ finishings are taken as those of the base to which the finish is applied. Tiling, on the other hand, is measured on the exposed face of the finishing. In practice this makes little difference as normally the structural room sizes are taken for the measurement of finishings and decoration. One would have to be careful, however, when measuring tiling to columns etc. as the face area of the tiling would be greater than the structural face area. Generally no deduction is made for voids within the area of the work when not exceeding 0.5 m^2. Rounded internal and external angles exceeding 10 mm, not exceeding 100 mm radius to in situ finishings are measured as linear items. When the radius exceeds 100 mm the work becomes curved. In the case of tiling these angles are measured only if special tiles are used.

CEILING FINISHINGS

Ceiling areas are taken from wall to wall in each room using the figured structural dimensions. The dimensions are usually best taken over the beams and the adjustments made later. If the majority of ceiling finishings are identical it may be advantageous to measure the ceiling over all the internal walls and partitions. From this measurement would be deducted the total lengths of the internal walls and partitions, possibly obtained from previous dimensions, squared by their respective thicknesses. Treating the surface of concrete by mechanical means or as it is traditionally known 'hacking concrete' to receive plaster, is measured superficially and is conveniently 'anded on' to the finishings dimensions. The application of bonding agents or other preparatory work is described with the plasterwork. Plasterboard or lath is also measured with the plaster finish.

WALL FINISHINGS

In the case of wall finishings it is best to total on 'waste' the perimeters of all rooms having the same height and finish. The areas of the walls may thus be contained in two or three dimensions instead of the large number there would be if each room were measured separately. As in the case of the measurement of walls, openings are generally ignored, the wall finish being measured across. Deductions for the work measured across openings together with work to reveals would be taken later with the measurement of windows, doors and blank openings. If large openings or infil panels occur from wall to ceiling height it is sometimes preferable to measure the finishings net. One must be careful to inform other measurers which openings have been measured net to avoid the possibility of a double deduction. As a general rule, if the walling has been measured across an opening, so would be the finishings.

If in situ wall finishings are applied to brick or block walls, then the raking out of the joints to form a key is deemed included. Preparation of concrete walls is dealt with as mentioned for ceilings above.

ANGLE BEADS ETC.

Metal angle, casing and movement beads are measured as linear items. Sizes are given in the description but working finishings to beads is deemed included.

DECORATION

The decoration to ceilings and walls should be measured with the plaster or other finish. If the decoration is all the same this can be 'anded on' to the plaster description. If, however, there is a mixture of several types of decoration, such as emulsion, gloss paint and wallpaper, it will probably be best to sub–divide the plaster measurement so that the appropriate decoration can be 'anded on'. Where all is, say, emulsion except wallpaper in one or two rooms it is usually more satisfactory to measure the whole as emulsion in the first instance and then to deduct the emulsion and add wallpaper for the appropriate area. Isolated paintwork where the girth does not exceed 300 mm or where the area does not exceed 0.5 m^2 is measured linear or enumerated respectively. Paintwork to walls, ceilings, beams and columns is described as work to general surfaces and if the same can be added together in the bill. Work to ceilings and beams exceeding 3.5 m above floor level has, however, to be so described in stages of 1.5 m. Papering has to be described in the two categories of ceilings with beams and walls with columns, the same rule regarding high ceilings applying.

CORNICES AND COVES

The measurement of cornices and coves is usually straightforward, as they normally run all round the room without interruption. Both in situ and prefabricated types are measured the length in contact with the base, the length already calculated for the perimeter of the walls being used. The measurement must include returns to projections and to beams if these occur in the ceiling. Ends, angles, and intersections are enumerated as extra over the items. Paintwork to cornices does not have to be kept separately if it is the same as that to the walls or ceilings. In fact it is questionable whether any adjustment should be made to the wall and ceiling decoration if this is the same, particularly if the cornice is contoured.

SKIRTINGS

There is a difference of opinion as to whether skirtings should be measured net or measured gross across openings. The disadvantage with measuring net is that the taker–off measuring finishings will have to ascertain the width of openings and possibly the depth of reveals. On the other hand wall decoration, as opposed to plaster, should be deducted behind the skirting. Thus, if the skirtings are measured gross with the deduction of decoration behind, when the doors are measured a deduction of plaster and decoration is usually made for the height of the opening. This results in the decoration being deducted twice behind the skirting and therefore an addition of decoration will have to be made for the length by the height of the skirting across the door opening.

Timber skirtings are measured linear, usually using structural wall face dimensions, but care will have to be exercised with measurements to isolated piers etc. as the length will be increased. The size and shape of the skirting and method of fixing if specified is given in the description but mitres, ends and intersections are deemed included except when in hardwood over 0.003 m^2 sectional area. If the skirting is fixed to a ground no deduction of plaster is made for the ground. Chair, dado and picture rails, rarely specified nowadays except in restoration work, are measured in the same way as skirtings.

In situ and tile skirtings are measured as linear items stating the height and thickness in the description. Fair and rounded edges are deemed to be included as are ends and angles.

WALL TILING

Firstly it is necessary to ascertain whether the wall tiling is fixed with adhesive or, more traditionally, bedded in mortar. The specification may call, particularly in the case of a tiled dado, for the wall behind the tiling to be plastered as the remainder of the wall and the tiles to be fixed to the plaster with adhesive. Alternatively there may be a cement and sand screed behind the tiling and the tiles either bedded in mortar or fixed with adhesive. If there is a mixture of wall tiling and plaster finish on a wall it is usually necessary for the back of the tiles to be flush with the face of the plaster. If special tiles with rounded edges are specified at top edges and external angles these are measured as linear items extra over the main work. The rules for measurement of tiling follow closely those for in situ finishings but attention is drawn to the paragraph headed generally above relating to measurements.

FLOOR FINISHINGS

The measurement of these has been covered in the chapter on floors and a decision has to be made as to where these are measured most conveniently.

STAIRCASE AREAS

Generally work in staircase areas has to be kept separately and the rules relating to the height of the work do not apply. When staircases are contained within lobby areas there is little difficulty in deciding what is and what is not a staircase area. When they are not it is not always clear and in this case the surveyor has to use his or her discretion and decide the sub–division. This decision should be passed on to the estimator by way of the bill so that later there is no argument.

DRY WALL LININGS

These are considered to be beyond the scope of this book, but briefly, open spaced grounds over 300 mm wide are measured superficial stating their size, spacing and fixing. Plasterboard dry linings are measured linear stating the height in stages of 300 mm in the description. Internal and external angles are measured as linear items.

Taking-off List	**SMM Reference**
Note: Generally work in staircase areas and plant rooms is kept separately.	
Ceiling finishings: (work over 3.5 m high kept separately)	
Plaster to concrete including attached beams	M20.2
Ditto to isolated and attached beams with different finish	M20.3
Hacking concrete (soffits)	E41.4
Decoration	M60.1
Plasterboard and skim coat	M20.2
Decoration	M60.1
Board as self finish	K10.2
Decoration	M60.1
Expanded metal lathing	M30.3
Plaster to ditto	M20.2
Decoration	M60.1
Pine boarding	K20.3
Decoration	M60.1
Wall finishings:	
Plaster to brick or block including attached columns.	M20.1
Ditto to isolated & attached columns with different finish	M20.4
Decoration	M60.1
Metal angle beads etc.	M20.24.8
Wall tiling	M40.1
Screed for wall tiling	M10.1
Plaster to concrete	M20.1
Hacking concrete	E41.4
Decoration	M60.1
Dry linings	K10.2
Grounds for ditto	G20.11
Decoration	M60.1
Coves and cornices	M20.17/19
Angles, ends etc. to ditto	M23.1–4
Decoration	M60.1
Skirtings (Timber)	P20.1
Decoration	M60.1
Floor finishings:	
Floor finish (Tile)	M40.5
	M50.5
Screeds	M10.5
Dividing strips	M40.16.4
	M50.13.5

EXAMPLE 11
INTERNAL FINISHINGS

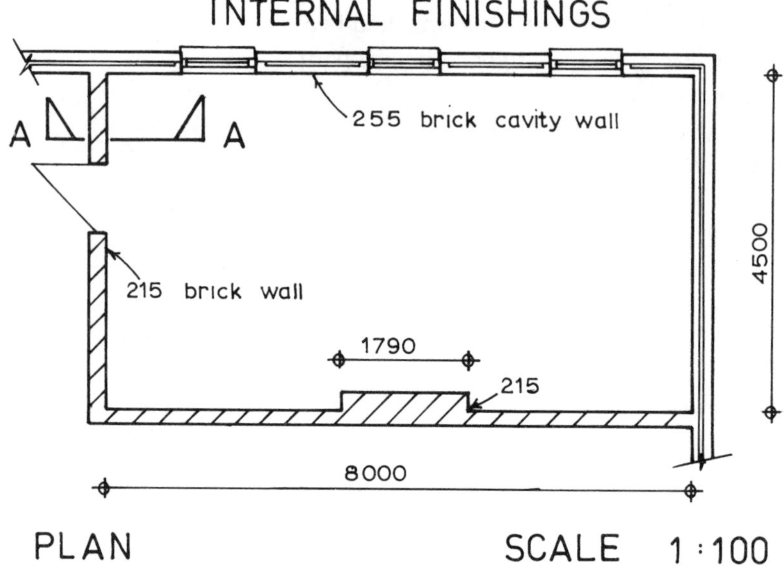

PLAN SCALE 1:100

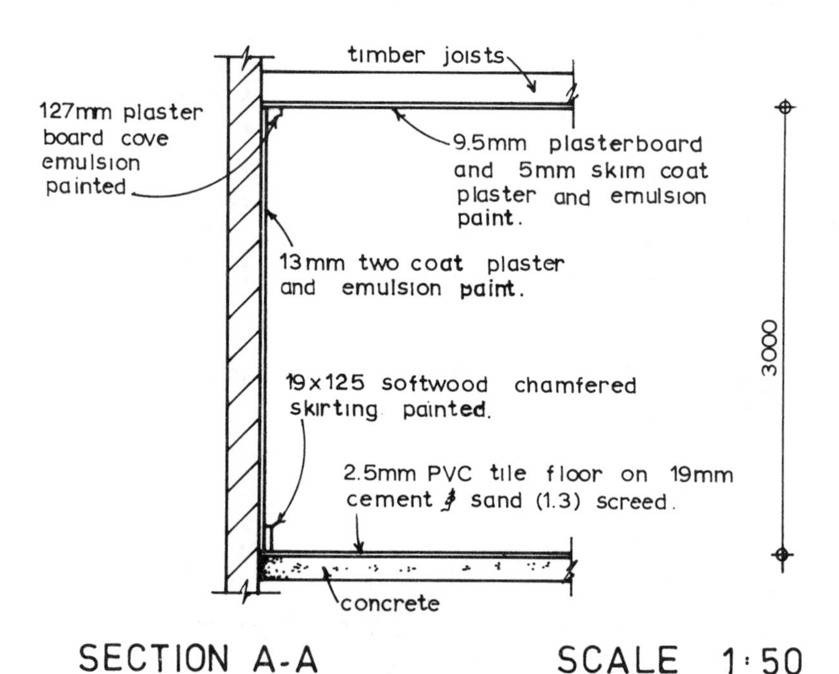

SECTION A-A SCALE 1:50

FIGURE 31

Dim		Description	Notes
		Internal Finishings 1.	
8.00 4.50		Plastered ceilings, width exceeding 300, 9.5 gypsum plasterboard fixed with rustproofed nails to softwood, 5 one coat Thistle board finish plaster &	M20.2.1.2 Reinforcing scrim to plasterboard deemed included M20.C3 Note that it may be necessary to measure noggings between the timber joists for fixing the edge of the plasterboard.
		Two coats emulsion paint general plaster surfaces, girth exceeding 300 &	M60.1.0.1
		PVC tiles 225 x 225 x 2.5, width exceeding 300, level, bedded with bitumen adhesive to cement screed &	M50.5.1.1 The floor is 'anded-on' to the ceiling in this case as the areas are identical.
		Cement & Sand (1:4) screed to floors, level, 19 thick in one coat to concrete base	M10.5.1 An item of "trowelling surface of unset concrete to receive paving" should be taken if not measured with the slabs. (E41.3)
1.79 0.22		Deduct [Pier	Deduction for area displaced by pier. Note that the minimum deduction rule does not apply as the void is at the perimeter. GR3.4

		Internal Finishings 2.	
		8000 4500 2/12500 = 25 000 2/215 = 430 25430	
25.00 3.00		Plastered walls, width exceeding 300, two coats 13 thick to brickwork [Pier	M20.1.1.1. Raking out joints of brickwork to form key deemed included F10. C1 (d)
2/ 3.00		Ditto width not exceeding 300	M20.1.2.1 Plasterwork to sides of the pier.
25.43 3.00		Two coats emulsion paint a.b.	M60.1.1.1 The width not exceeding 300 does not apply to this paintwork as the surface is not isolated.
2/ 3.00		Accessories, galvanised perforated metal angle bead, vertical, with 50 returns, fixed to brickwork with masonry nails	M20.24 Working finishings to beads is deemed included. Note that external and internal angles without angle beads are deemed included (C4)

		Internal Finishings 3.		
			[Coves	
	25.43		Gyproc cove 127 girth fixed with adhesive	M20.17 No adjustment has been made for the emulsion paint as this is regarded as equal to that displaced on the walls and ceilings.
6/	1		Extra over ditto for internal angles	M20.23.2
3/	1		Ditto for external angles	M20.23.3
			[Skirting	
	25.43		Wrot softwood skirting 19 x 125 chamfered plugged & screwed to brickwork	P20.1.1 Mitres etc. to skirting deemed included P20 C1 State if size is finished. Method of fixing described is not at contractor's discretion.
			&	
			Knot, prime, stop & three coats gloss paint on general surfaces woodwork, isolated surfaces not exceeding 300 girth	M60.1.0.2

			Internal Finishings 4.	
	25.43		Prime only general surfaces woodwork, isolated surfaces not exceeding 300mm, application on site prior to fixing & Deduct Two coats emulsion paint general plaster surfaces x 0.13 = m²	M60.1.0.2.4 Priming to the back of the skirtings before fixing.
	To TAKE		Adjustment of finishings for openings.	Normally the adjustment of finishings for door,window, fireplace etc. openings is left for adjustment with the measurement of these items.

12 Windows

SCHEDULES

As with internal finishings measurement and for similar reasons, usually it is prudent to prepare a schedule of windows before commencing measurement. An example of a typical schedule is given in Appendix 2 but the headings may have to be amended to suit the particular circumstances. Before preparing the schedule the windows should be lettered or numbered on the floor plans and the elevations. If there are several floors to the building then a system of numbering should be devised which enables windows on a particular floor to be located readily. For example, a letter could be allocated to each floor followed by a number for each window, the numbering starting at a particular point on each floor and proceeding, say, in a clockwise direction round the building. Whilst windows are usually scheduled floor by floor there may be good reason to work by elevations or by window types or even by a combination of all three. The total number of windows entered on the schedule should be checked carefully with the total number shown on the drawings to ensure that none is missed. A further check must be made to ensure that the windows shown on the elevations tie–up with those shown on the plans. Discrepancies may occur because clerestory windows or windows to mezzanine floors are sometimes shown on the elevations but not on the plans. Any differences found should be mentioned to the architect and the matter resolved before commencing measurement. The schedule should aim to set out the details of each window so that it is hardly necessary to refer to the drawings during measurement. Usually it should be possible to measure together in one group all windows of one type irrespective of their size, thickness of wall etc.

A note should be made at the commencement of the measurement for each group of the numbers of windows being dealt with and care being taken throughout that this total is accounted for in each item. A common fault with beginners is to separate the entire measurement of windows of the same type which are of different sizes or in different thicknesses of walls. Such a method may give less trouble but probably will take longer. Although the grouping of windows of different sizes may require more concentration and care, proper use of the schedule will considerably simplify the work and, additionally, the amount of sorting and collection of cut and shuffle slips or abstracting will be reduced.

SUBDIVISION

The measurement of each type of window may be divided into the following subdivisions:

(a) Windows
  (1) timber, metal or upvc casement and fixing
  (2) glass (if not included in (1) above)
  (3) ironmongery (ditto)
  (4) painting (if required)
(b) Opening adjustments
  (1) deduction of brickwork and external and internal finishings
  (2) support to the work above the window
  (3) damproofing and finishings to the head externally and internally
  (4) ditto to the external and internal reveals
  (5) ditto to the external and internal sills

If these items are followed through systematically in each group there will be less chance of items being missed.

TIMESING

Window dimensions will probably contain a fair amount of 'timesing' and great care is necessary to ensure that this is done correctly. As each of the sub–divisions of measurement is completed it is advisable to total the timesing of each item to ensure that it equals the number of windows being measured. If, for instance, twenty windows are being measured in a group, timesing of deductions, lintels, sills etc. should total twenty unless a change in specification or design for a particular window requires a smaller number.

SPECIAL FEATURES

Usually any special features such as small canopies over or decorative brickwork under which definitely relate to the window will be measured with the window.

DORMER WINDOWS

In the case of dormer windows the window itself as indicated in sub–division (a) above may be taken with the other windows whilst the adjustment of the roof would normally be taken with the roofs. This division will generally be found to be convenient, particularly if the roofs and windows are being measured by different persons. The opening for a dormer window may sometimes be partly in the wall and partly in the roof, in which case the wall adjustment would be made with the window measurement, in the same way as the other window openings.

ADJUSTMENTS

When measuring a group of items, the advantage of taking initially the same description for similar work and then making an adjustment for small differences often becomes evident in the measurement of windows. For example if all the windows are glazed in clear glass except for a small proportion which have patterned glass, the simplest way of measuring may be to take all as clear glass and then, checking over carefully with the schedule, making adjustment for those which need patterned glass. If, as is not impossible, the measurement of the patterned glass to one or more windows should be missed, at least if clear glass has been measured to them all the error will be much less than if none had been measured. Similarly when making adjustments for the openings, different decoration may be applied to the walls. If the predominating one is chosen for the deductions to all windows then, as an adjustment, the true decoration may be deducted where appropriate and the predominating finish, deducted earlier, added back. Apart from minimising the effect of errors this method also facilitates the grouping of descriptions under the same measurements.

WINDOWS

Timber, metal and plastic windows and frames are enumerated and described with a dimensioned diagram. A reference to a catalogue or standard specification may remove the need to provide a diagram. The method of fixing must be shown unless this is at the discretion of the contractor. Bedding and pointing of the frame is given as a linear item. Timber window boards and cover fillets are measured as linear items giving their cross section dimensions and labours in the description.

GLASS

The size of each pane of glass can sometimes be obtained from manufacturer's catalogues, otherwise it has to be calculated from the overall size of the window deducting for frames and mullions etc. Standard plain glass, when not exceeding 10 mm in thickness and when not exceeding 4 m^2 in area, is measured superficial. The size of the panes is indicated in the description as follows:

- not exceeding 0.15 m^2 in area (stating the number of panes)
- 0.15 to 4 m^2 in area

If there are more than fifty identical panes then their number and size is stated. Raking and curved cutting to glass is deemed to be included. Non–standard panes outside the thickness and size classification given above are enumerated the size being given in the description. If rebates for glass are over 20 mm then this is stated in the description the rebate depth being given in 10 mm stages. Adjacent panes of glass required to align with each other have to be so described. One of the rare occasions when waste is allowed for in measurement occurs in glazing in that panes of irregular shape are measured the smallest rectangle from which the pane can be obtained. Bedding the edges of glass in strips or channels is measured as a linear item.

If the glazing is double then this is stated and the glazed area is doubled for calculating the bill quantity. If hermetically sealed double glazing units or special glasses are specified then the panes are enumerated stating their size.

IRONMONGERY

Each item of ironmongery, if not supplied with the window, is enumerated and described giving the nature of the background to which it is fixed. The description of ironmongery is simplified by reference to manufacturer's catalogues, being careful to give sizes, material and finish if alternatives are listed. Frequently it is convenient to include a PC sum for the supply of ironmongery and measure out the fixing. This avoids the necessity to give a full description for the ironmongery which is impossible at the measurement stage if a selection has not been made.

DECORATION

Painting is measured as a superficial item the measurement being taken over frames, mullions, transoms, sills and glass. The description includes the size of the panes, averaged if of more than one size, classified as follows:

- not exceeding 0.1 m^2
- 0.1 to 0.5 m^2
- 0.5 to 1 m^2
- exceeding 1 m^2

The measurement is deemed to include such items as paint on opening edges and the consequential extra frame, cutting in and work on glazing beads. Painting which is external has to be so described. Priming only to backs of frames before fixing is measured as a linear item if not exceeding 300 mm girth.

OPENINGS

Unless the opening has rebated reveals the same dimension can be used for the deduction of the wall, cavity and external and internal finishings. After having made this adjustment it is best to consider the perimeter of the opening in the order of head, jambs and sill. Precast concrete lintels are enumerated; the size, shape and reinforcement being included in the description. In situ concrete lintels are measured as isolated beams and the cubic measurement of the concrete, and the measurement of the formwork and reinforcement each have to be taken separately. Brickwork displaced by lintels is only deducted for height to the extent of full courses displaced and for depth into the wall to the extent of full half brick beds displaced. Damp proof courses to heads forming cavity trays are so described. Proprietary steel lintels are enumerated giving the manufacturer's reference. Facework to arches is measured as a linear item stating the number. Plaster to the reveal, if not exceeding 300 mm wide, is measured as a linear item. Decoration to the reveal is measured as the work to the adjacent wall. At the jambs cavity closing is measured as a linear item, facework to the reveals is deemed included but vertical damp proof courses are measured. Finish to the reveals is measured as for the head of the opening. Precast concrete sills are measured in the same way as lintels with the same rules for adjustments. Facework to sills is measured as a linear item stating if set weathering and the dimensions.

Taking–off List	**SMM Reference**
Windows – timber, plastic or metal (including fixing)	L10/11/12.1
Bedding and pointing frames	L10.8/9/10
Window board	P20.4
Cover fillets	P20.2
Ironmongery	P21.1
Glazing	L40.1
Decoration	M60.2
Opening:	
Deduction of: wall	
cavity	as before
plaster and decoration	
Head: Lintel – concrete	F31.1
proprietary	F30.16
Arch	F10.6
Cavity tray	F30.2
Adjustment of wall	as before
Finish to reveal – plaster	M20.1
decoration	M60.1
Metal angle beads	M20.24.8
Jamb: Closing cavity	F10.12
Damp proof course	F30.2
Finish to reveal	as above
Sill: Concrete and stoolings	F31.1/2
Brick	F10.15
Tile	M40.7
Damp proof course	F30.2
Water bar	P21.1
Adjustment of wall	as before

EXAMPLE 12
WINDOW

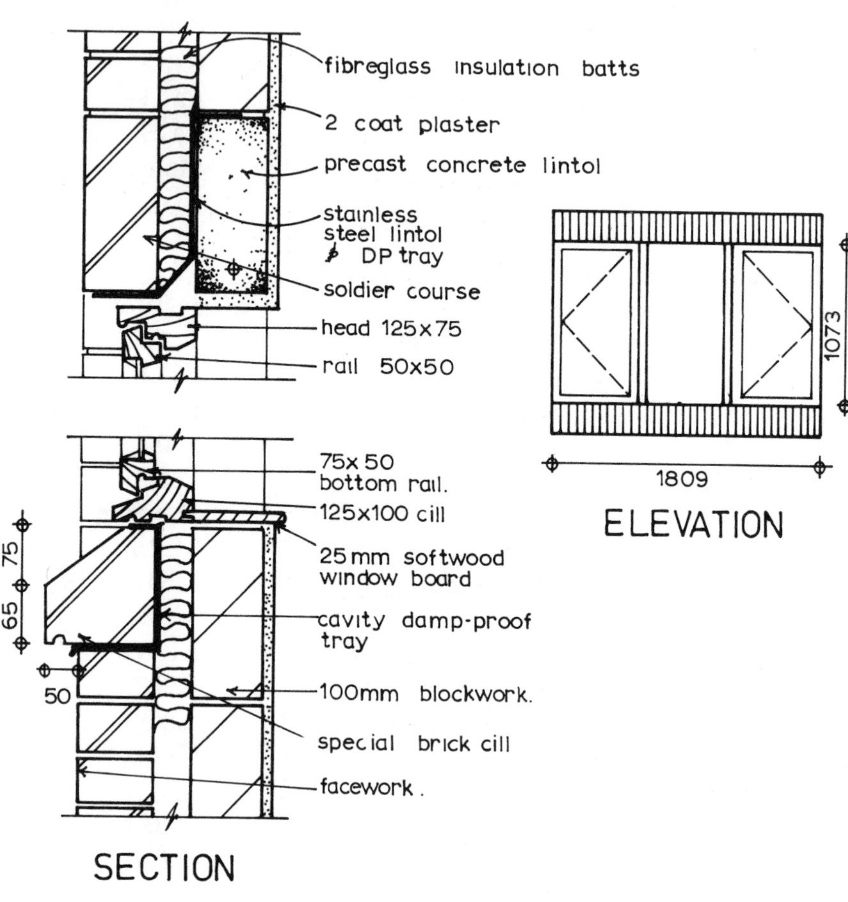

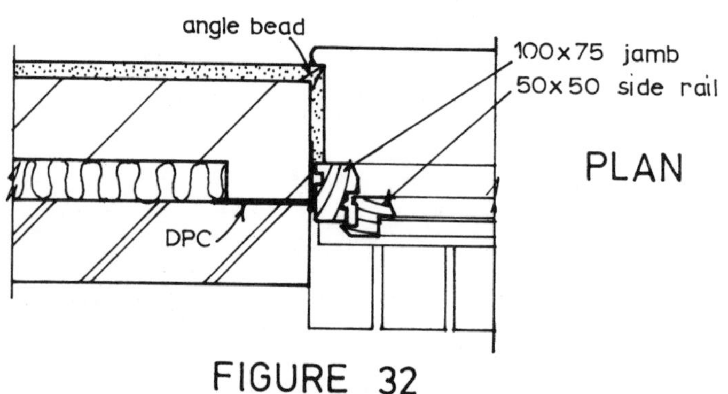

FIGURE 32

Windows 1.

Timesing	Dimension	Squaring	Description	Notes
	1		Wrot softwood window & frame as dimensioned diagram Fig. 32, size 1809 × 1073 overall with one fixed and two opening lights, the frame fixed with four galvanised mild steel fixing cramps screwed to back of frame and built into blockwork	L10.1.0.1 The window is described with the assistance of a dimensioned diagram. Alternatively the description could refer to a manufacturers catalogue reference.
2/	1.81		Bed wood frames in gauged mortar (1:1:6) and point one side in mastic	L10.10.
2/	1.07			
	1.81		Wrot softwood window board 25 × 150, rebated and rounded one edge, screwed & pellated to masonry	P20.4.1. Ends deemed included.

		Windows 2.	
7/	1	Pairs 75 pressed steel butts to Softwood & Brass casement stay & two pins as Messrs. ______ Ref. ______ to Softwood & Brass casement fastener as Messrs. ______ Ref. ______ to Softwood	P2.1.1.1. If the ironmongery has not been selected at this stage a p.c. sum would be included to cover the supply of ironmongery with provision for the addition of profit by the tenderers. The fixing of each item of ironmongery would be enumerated. Proprietary windows are often delivered with the ironmongery already fixed at the factory, in which case the ironmongery required would be included in the description of the window.
		[Glass. Width 1809 Frame. 2/40 = 80 Mullions. 2/30 = 60 140 ÷3) 1669 Width fixed. = 556 Side rails. 2/20 = 40 Width opg. = 516	Calculation to ascertain size of glass to the fixed and opening lights. Manufacturers catalogues frequently contain glass sizes for various window types.

Windows 3.

		Height.
		1073
Head.	40	
Sill.	50	90
Height fixed =		983
Top rl.	20	
Btm rl.	30	50
Height opg. =		933

Timesing	Dimensions	Description	Notes
2/	0.52 0.93 0.56 0.98	Standard plain glass, 3 clear sheet (ordinary quality), glazing with sprigs & putty to wood in panes area 0.15 - 4 m²	L40.1.1.2
2/	1.81 1.07	Knot & prime only glazed wood windows, panes area 0.50 - 1.00m², girth exceeding 300, application on site prior to fixing	M60.2.3.1.5. The whole window is to be primed on site before fixing. The pane size is averaged for the whole unit.

		Windows 4.		
	1.81 1.07		Two undercoats & one finishing coat gloss paint on primed glazed wood windows, panes area 0.50 - 1.00m², girth exceeding 300 & Ditto externally	M60.2.3.1 The painting is measured over the glass and work to opening edges and cutting in to glass is deemed included.
			[Wdw. board.	
	1.81 0.35		Knot & prime only general surfaces woodwork, application on site prior to fixing	M60.1.0.1.4
	1		Two undercoats & one finishing coat gloss paint on primed general surfaces woodwork isolated areas not exceeding 0.50 m²	M60.1.0.3.

Windows 5.

[Opening adjustment.

1.81		Deduct Wall in facings half brick thickness, vertical a.b.	When making deductions to create the opening for the window the descriptions need only be long enough to identify the item.
1.07		&	
		Deduct Forming cavity in hollow wall 50 wide including ties & fibreglass insulation batts a.b.	
		&	
		Deduct Block wall 100 thick a.b.	
		&	
		Deduct Plaster walls width exceeding 300 to blockwork a.b.	
		&	
		Deduct Two coats emulsion paint plastered walls a.b.	

Windows 6.

[Head.
[Lintol.

1809
2/150 : 300
2109

1		Precast concrete (27 N/mm² - 20 agg.) lintol, rectangular section 2110 × 100 × 210, reinforced with one 16 diameter mild steel bar and build in to blockwork in gauged mortar (1:1:6)	F31.1.1.
		&	
		3 × 350 girth stainless steel combined lintol & cavity tray 2110 long & build in to blockwork in gauged mortar (1:1:6)	F30.16.1

Windows 7.

Dimensions	Description	Notes
1.81	Flat soldier arch 215 wide on face in facing bricks in gauged mortar (1:1:6) half brick thick, width of exposed soffit 20, including pointing (In Nr. 1)	F10.6.1
2.11 0.21	Deduct Block wall 100 thick a.b.	The block wall is deducted for the lintel as one complete course is displaced.
	&	
	Deduct Plaster walls width exceeding 300 to blockwork a.b.	The plaster for the area of the lintel is deducted assuming that a different specification is used for plastering to concrete.
2.11 1.81	Plaster walls width not exceeding 300, two coats, 13 thick to concrete including bonding agent	M20.1.2.1. Plastering to the exposed face and soffit of the concrete lintel.
1.81	Galvanised perforated metal angle bead horizontal with 50 returns fixed to concrete with plaster dabs	M20.24

Timesing	Dimension	Squaring	Description	Notes
			Windows 8.	
	1.81 0.10		Two coats emulsion paint plastered walls a.b.	M60.1.0.1 This is not an isolated surface and therefore is not classified as not exceeding 300 girth.
	1.81 0.23		Deduct Wall in facings half brick thick a.b.	Deduction of brickwork displaced by arch.
			[Jambs.	
2/	1.07		Closing cavities 50 wide with 100 blockwork, vertical	F10.12.1.1
2/	0.10 1.07		Damp proof course width not exceeding 225, vertical, of Hyload pitch polymer bedded in gauged mortar (1:1:6)	F30.2.1.1

Timesing	Dimension	Squaring	Description	Reference
			Windows 9.	
2/	1.07		Plaster walls width not exceeding 300, two coats 13 thick to blockwork	M20.1.2.1
			&	
			Galvanised perforated metal angle bead vertical with 50 returns fixed to blockwork with masonry nails	M20.24.8
			&	
			Two coats emulsion paint plastered walls a.b.	M60.1.0.1
			× 0.10 = m²	
			[Sills.	
	1.81		Facework sill, horizontal, purpose made splayed bricks 65 × 150 × 140 throated and set projecting 50 and pointing to exposed faces	F10.15.1.3

		Windows 10.		
1.81 0.15		**Deduct** Wall in facings half brick thick a.b.		
		2/. 1809 300 2109		
2.11 0.25		Damp proof course width exceeding 225, horizontal, of Hyload pitch polymer bedded in gauged mortar (1:1:6)		F30.2.2.3 The damp proof course is extended beyond the jambs to give extra protection.

13 Doors

SUBDIVISION

The measurement of doors is conveniently divided as follows:

(a) external doors
(b) internal doors
(c) blank openings

Whilst it is not essential to take external and internal doors separately it will probably be found that there are different types of doors with different finishes and therefore they have little in common. The measurement of each group can in turn be divided into similar sections to those given for windows in the previous chapter. One exception being that, for internal doors, in place of sills, flooring in the opening will have to be taken unless already measured with floors or finishings. Steps to external doors may be measured at this stage together with canopies or other special features in connection. Blank openings without windows or doors should form a separate sub–section which includes all places where walls and finishings have been measured over and have to be deducted. Adjustments for openings are made in the same way as for windows. Some surveyors measure all the windows and doors themselves and then create a completely separate section for opening adjustments. Probably it is better to deal with openings stage by stage whilst the sizes of the joinery items are in mind.

SCHEDULE

As with windows schedules should be compiled so that details of all the doors can be seen at a glance. An example of a typical door schedule is given in Appendix 2. Doors on the plans should be lettered or numbered serially as described for windows in the last chapter. External doors may be included with the window numbering and internal blank openings included with internal doors.

MEASUREMENT OF DOORS

Doors are enumerated and described with a dimensioned diagram. Frequently doors are of standard types from particular manufacturers and catalogue references could be given instead of a diagram. Care must be taken to describe the exact details when options are offered by the manufacturer, for example various face finishes, cell construction and lippings to flush doors. It is important to establish where any fire resisting doors are required and to describe these correctly and to consider if special frames are necessary. In the case of multi–leafed doors each leaf is measured as one door. Doors hung folding must have their rebated meeting stiles mentioned in the description and swing doors their rounded stiles. Weatherboards and glazing beads are also included in the description or shown in the diagram.

GLASS AND IRONMONGERY

The notes relating to the measurement of glass and ironmongery in the previous chapter on windows apply equally to doors. Invariably glass in doors is fixed with beads and fitted in a protective strip to guard against slamming effects.

FRAMES AND LININGS

Firstly it is necessary to understand the difference between frames and linings. Frames, as their name suggests, are framed together at angles, usually in the workshop, and delivered to the site as a unit. External doors and internal doors in thick walls invariably have frames. Linings, again as their name suggests, line the opening and are usually made on site and are, at the most, tongued together at angles. For both allowances in length should be made for jointing and in the case of frames for horns. The length of the frame or lining is best calculated from the size of the door, adding at each corner for the thickness of the timber and jointing. A complete frame or lining to a door is known as a set. The total number of sets is given in the bill stating how many are identical. The jambs and heads and, if required, the sills, mullions and transoms are each measured separately as linear items. In the case of sills, mullions and transoms the number of each must be given in the description. Within each measurement the number of identical sections must be given. The reason for this is that members of the same size, with the same labours and therefore having the same description, may vary in shape; for example the rebates may be of a different size or in different positions. Alternatively door frame and lining sets may be enumerated and described. This method of measurement would be appropriate if there are several which are identical. The fixing of frames and linings must be described if not at the discretion of the contractor.

ARCHITRAVES

Architraves are measured as linear items, ends and mitres being deemed included. If the architrave is shaped then the length measured should be that around the external edge as, where mitres are cut, a wedge of timber is lost. In some situations, for example where a return wall is adjacent to a door, it may not be possible to insert an architrave and a cover fillet would be used instead.

DOORS WITH SIDELIGHTS

Doors with sidelights or borrowed lights are measured as described in windows, the dimensioned diagram showing details of the unit.

PAINT

Painting on doors is measured as a superficial item and if the door is panelled has to be described as an irregular surface. In the case of panelled doors an extra allowance has to be made for the girth of mouldings, probably best done by multiplying the flat area by $1\frac{1}{8}$. As a matter of interest this method dates from days when measurements in square feet were divided by nine to convert to square yards for billing. Moulded surfaces were divided by eight to make an allowance for the extra area. Heavily moulded surfaces, unlikely to be found in modern construction, were divided by seven. Whilst this method is very approximate, painting is a comparatively cheap item and does not warrant detailed calculation. The thickness of the doors has to be added to the face area to include for the painting to edges. Painting to glazed doors is so described and measured as painting to windows. Painting to frames, linings and architraves is measured superficial if over 300 mm girth and linear if under 300 mm girth.

Taking-off List	**SMM Reference**
Doors	L20.1
Decoration to door	M60.1
Decoration to glazed door	M60.4
Ironmongery	P21.1
Glazing	L40.1
Bedding edge glass	L40.11
Frame/lining	L20.7
Bedding frame	L20.8/9/10
Architraves	P20.1
Cover fillets	P20.2
Decoration frames etc.	M60.1
Opening adjustment	See Windows
Adjust skirting	As before
Flooring in opening	As before
Brick threshold	F10.16
Brick step	F10.18
Precast concrete threshold	F31.1
Matwell	N10.1

EXAMPLE 13

INTERNAL DOOR

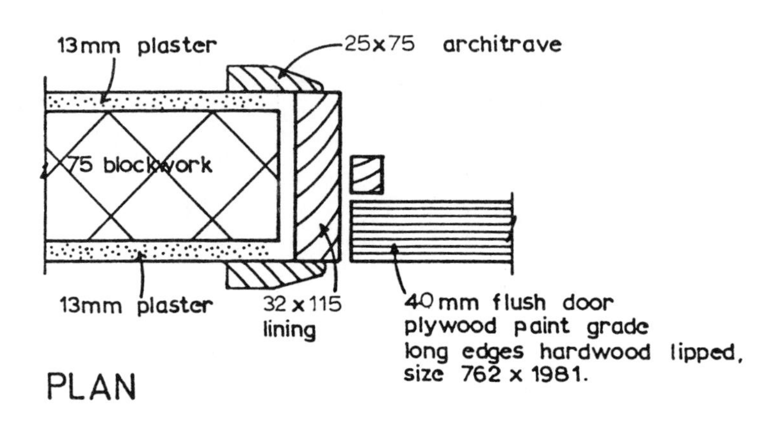

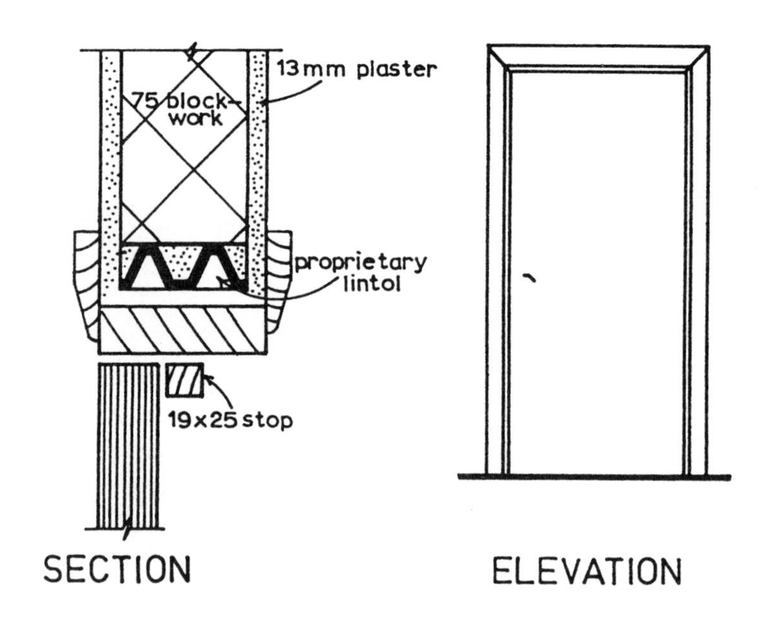

FIGURE 33

Internal Doors 1.

1		40 mm Flush door 762 x 1981 as dimensioned diagram (Figure 33)	L20.1.0.1 The description should be sufficient to identify the door readily and should include any information not included on the diagram.
		762 + 40 = 802; 1981 + 40 = 2021	
2/ 0.80 2.02		Knot, prime, stop two undercoats one finishing coat gloss paint general surfaces woodwork girth exceeding 300	M60.1.0.1 The thickness of the door is added to the face dimensions, which when twiced for both faces, covers the painting to all edges of the door if specified.
		P.C. of £______ for ironmongery	A52.1.1 Assuming that the ironmongery is to be supplied by a nominated firm.
		Add for profit	

		Internal Doors 2.		
	Fixing the following ironmongery to hardwood lipped softwood flush doors			P21.1.1 The nature of the background to which the ironmongery is fixed has to be stated.
1			Pair 75 pressed steel butt hinges	Sufficient information should be given in the descriptions to enable the fixing of the item to be priced.
1			Mortice lock & Silver anodised aluminium lever handle furniture	
		End of fixing		
			[Linings Head. 762 2/32 : 64 826 Jamb. 1981 ½/32 : 16 1997	

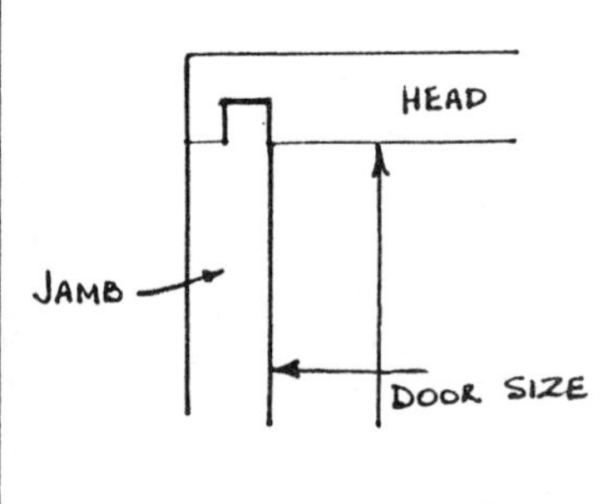

Internal Doors 3.

The following in Nr 1 door lining set tongued at angles

L20.7

The number of door lining sets has to be given. The heading would also include the number of identical sets.

0.83		Head 32 x 115 wrot softwood
2/ 2.00		Jamb 32 x 115 wrot softwood plugged & screwed to blockwork

Method of fixing is stated if not at the discretion of the Contractor.

End of door lining set

1981
− 19
1962

0.76		Stop 19 x 25 wrot softwood
2/ 1.96		

P20.2.1

DOOR SIZE

ELEVATION OF STOP

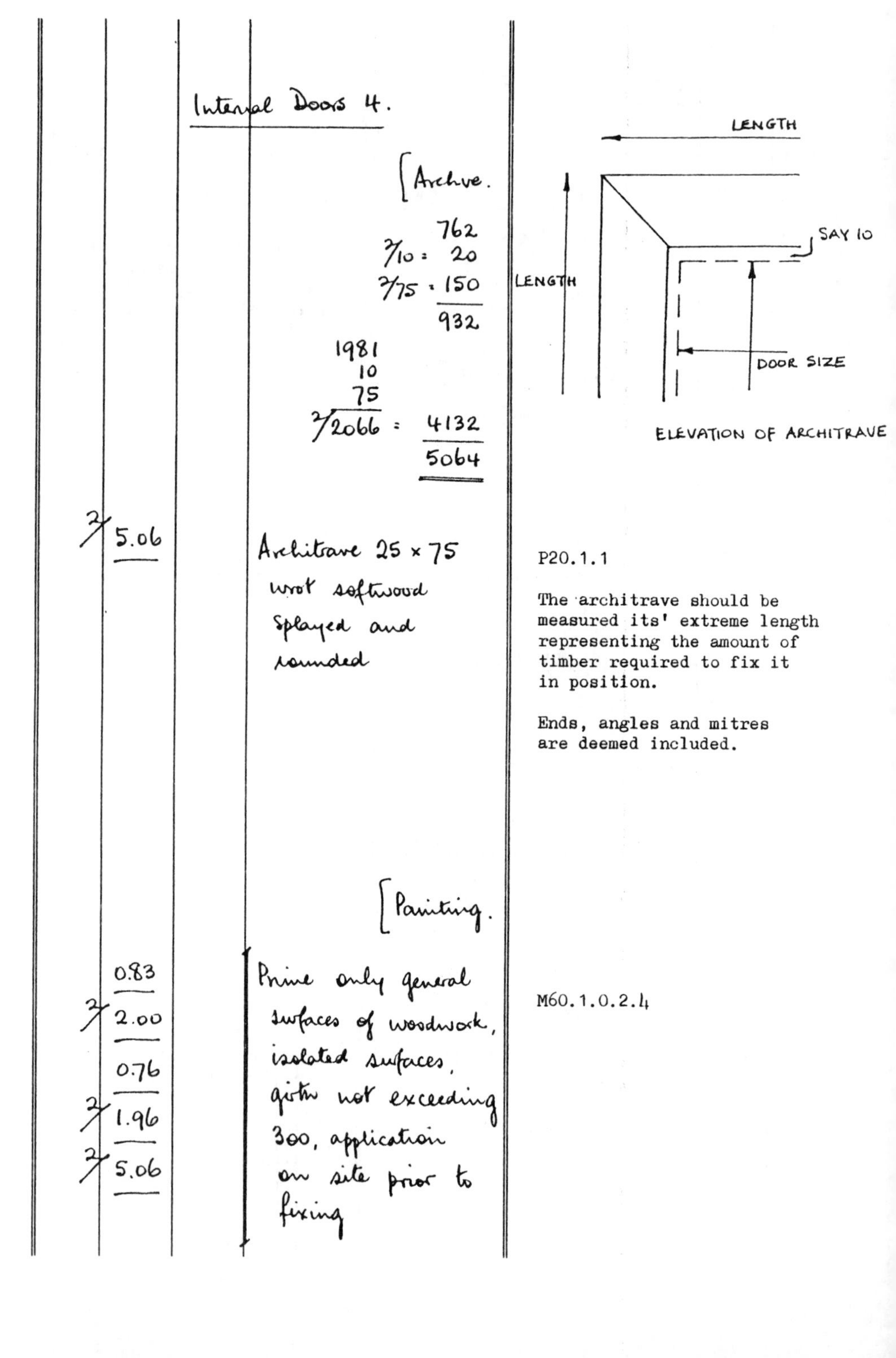

Internal Doors 4.

[Archve.

762
2/10 = 20
2/75 = 150
932

1981
10
75
2/2066 = 4132
5064

2/ 5.06 — Architrave 25 × 75 wrot softwood splayed and rounded

P20.1.1

The architrave should be measured its' extreme length representing the amount of timber required to fix it in position.

Ends, angles and mitres are deemed included.

[Painting.

0.83
2/ 2.00
0.76
2/ 1.96
2/ 5.06

Prime only general surfaces of woodwork, isolated surfaces, girth not exceeding 300, application on site prior to fixing

M60.1.0.2.4

Internal Doors 5.

Timesing	Dimensions	Squaring	Description	Side notes
			115 2/19 = 38 2/10 = 20 2/75 = 150 2/25 = 50 373	Calculation for girth of paint.
			826 2/1997 = 3994 4820	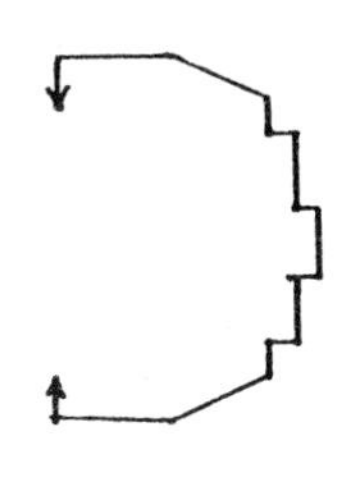
	4.82 0.37		Knot, prime, stop, two undercoats and one finishing coat gloss paint, general surfaces woodwork, girth exceeding 300	M60.1.0.1
			[Opening adjustment. 762 / 2/32 = 64 / 826 — 1981 / 32 / 2013	
	0.83 2.01		Deduct Block wall 75 thick a.b.	

Timesing	Dimension	Squaring	Description	Notes
			Internal Doors 6.	
2/	0.83 2.01		Deduct Plastered walls width exceeding 300, two coats a.b. & Deduct Two coats emulsion paint general plaster surfaces a.b.	The small amount of emulsion paint behind the architraves is not considered worth adjusting. Consideration should be given to this if the finish is more costly.
			[Lintol 826 brg. 2/75 = 150 976	Lintels usually available in 300 mm increments therefore state length as 1200
	1		Proprietary coated galvanised steel lintol type ______, 1200 long & building into blockwork	F30.16.1.1 No deduction of blockwork for lintel as full course of blocks not displaced.

			Internal Doors 7.	
2/	0.93		Deduct Wrot softwood skirting 19 x 125 a.b.	Adjustment of skirting necessary if skirtings were not measured net previously.
			&	
			Deduct Knot, prime, stop, two undercoats one finishing coat gloss paint general isolated surfaces woodwork not exceeding 300 a.b.	
			&	
			Deduct Prime only general isolated surfaces woodwork girth not exceeding 300 a.b.	
			&	
			Add Two coats emulsion paint general plaster surfaces a.b..	Addition for emulsion paint which has been deducted twice previously, once when the skirting was measured and once when the door opening adjusted.
			x 0.13 = m^2	
		To TAKE:	Flooring in opening	Flooring in opening to be measured if not previously taken.

EXAMPLE 14

EXTERNAL DOOR

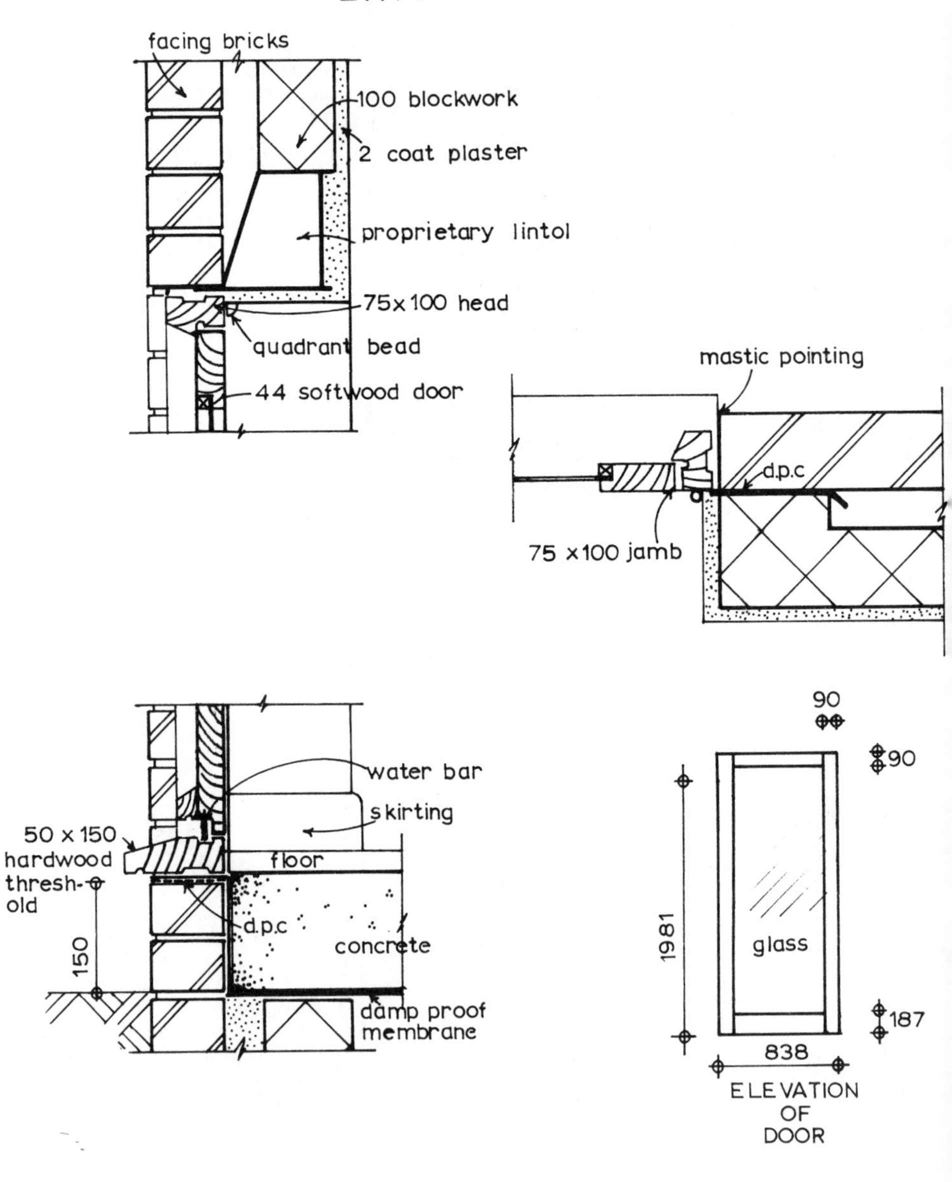

FIGURE 34

External Doors 1.

1	44 mm Wrot softwood single panel glazed door 838 x 1981 as dimensioned diagram (Figure 34) with weatherboard and 16 x 25 rounded hardwood glazing beads 838 + 44 = 882 1981 + 44 = 2025 Weatherbd. 2/50 = 100 2125	L20.1.0.1 The description should make it clear if items such as weatherboards and glazing beads are included.
0.88 2.03	Knot, prime, stop, two undercoats one finishing coat gloss paint general surfaces woodwork girth exceeding 300	M60.1.0.1 Painting is deemed to be internal unless otherwise described.
0.88 2.13	Ditto externally	

External Doors 2.

[Ironmongery.

	1		Pair 100mm Cast iron butt hinges, fixing to softwood	P21.1.1 If the ironmongery can be referenced to a manufacturer's catalogue or British Standard then there is no need to include a P.C. Sum. The nature of the background to which the item is fixed has to be stated.
	1		Brass cylinder nightlatch as Messrs. ________ Nr ____, fixing to softwood	
	1		Letter plate as Messrs. ________ Nr ____, 267 × 79, silver anodised aluminium, fixing to softwood including perforation	

External Doors 3.

[Glazing.

	838	100	1981
2/100 :	200	200	300
	638		1681
Rebates 2/16 :	32		32
	670		1713

1		Special glass, 6.4 laminated, clear float quality, size 670 x 1713, glazing to wood with screwed beads (measured)	L40.3.1

[Frames.

Jambs.	1981
Head	50
Cill	50
	2081

Head/Cill	838
Jambs 2/.	100
	938
Horns 2/75 =	150
	1088

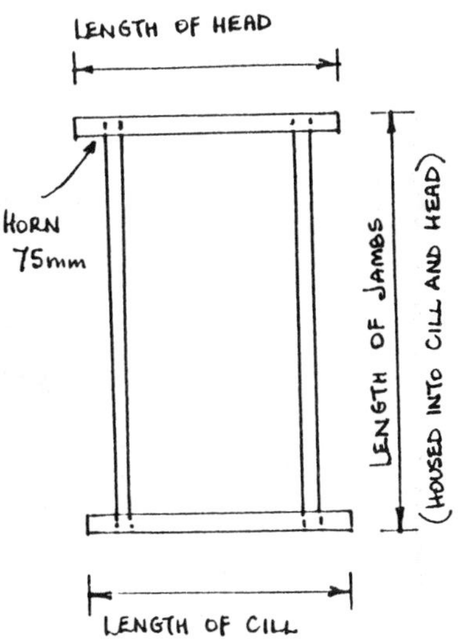

The following in Nr 1 door frame set

L20.7

The heading would also indicate the number of identical sets.

External Doors 4.

	1.09	Head 75 x 100 wrot softwood, splayed, rebated, check throated and grooved	L20.7.2.1
		&	
		Sill 50 x 150 wrot hardwood, weathered, twice grooved, throated and rounded (in Nr 1)	L20.7.3.1 The number of sills has to be given as some sets may not have sills.
2/	2.08	Jamb 75 x 100 wrot softwood, splayed, rebated, check throated and grooved, fixed to brickwork with galvanised frame cramps	L20.7.1.1 Method of fixing has to be stated where not at discretion of the Contractor. In this case the requirement is for the use of cramps, the specification of the cramp has been left to the Contractor.

End of door frame set

Timesing	Dimensions	Squaring	Description	Reference
			External Doors 5.	
2/ 2/	2.08 0.94		Bed wood frames in gauged mortar (1:1:6), including pointing in mastic	L20.10
	1		6 × 40 × 938 long galvanised water bar bedded in mastic to groove in hardwood cill	P21.1.1
			[Quadrant bead.	
2/	0.94 2.08		16 Wrot softwood quadrant bead	P20.2.1

External Doors 6.

[Decoration.

2/ 0.94		Prime only general surfaces of woodwork, isolated surfaces, girth not exceeding 300, application on site before fixing	M60.1.0.2.4
2/ 2.08			
		&	
		Knot, prime, stop, two undercoats and one finishing coat gloss paint, general isolated surfaces woodwork girth not exceeding 300	M60.1.0.2
		&	
		Ditto externally	

		External Doors 7.	
			[Opening adjustment.
0.94		Deduct Walls in facings half brick thickness a.b.	
2.08		&	
		Deduct Walls in Concrete blocks 100mm thickness a.b.	
		&	
		Deduct Forming cavities in hollow wall 50 wide a.b.	
		&	
		Deduct Plastered walls, exceeding 300, two coats a.b.	
		&	
		Deduct Two coats emulsion paint general plaster surfaces a.b.	

The deduction of the inner skin and cavity may have to be increased in height for the thickness of the concrete ground slab. (See note at end of example).

		External Doors 8.		
			{Head.	
			938 Brg. 2/150 = 300 1238	F30.16.1.1
	1		Proprietary coated galvanised steel lintel type____ 1350 long and building into brickwork	Lintels usually available in 150 increments therefore state length as 1350 No deduction of blockwork for lintel as full block not displaced.
	1.35 0.15		Deduct Plastered walls, exceeding 300, two coats ab.	Deduction of plaster to face of lintel.
	1.35		Plastered walls, width not exceeding 300, two coats 13 thick to metal lathing base including dubbing 10 thick	M20.1.2.1 Addition of plaster to recessed expanded metal face of lintel.

		External Doors 9.		
	0.94		Plastered walls, width not exceeding 300, two coats 13 thick to metal lathing base	M20.1.2.1 Plaster to the reveal at the head of the window.
			&	
			External angle bead to ditto	M20.24.8
			&	
			Two coats emulsion paint general plaster surfaces a.b. x 0.15 = m²	M60.1.0.1 Not described as "not exceeding 300" as surface is not isolated.
			[Jambs.	
2/	2.08		Closing cavities 50 wide with 100 blockwork vertical	F10.12.1.1
2/	2.08 0.10		Damp proof course width not exceeding 225 vertical of Hyload pitch polymer bedded in gauged mortar	F30.2.1.1

			External Doors 10.	
2/	2.08		Plastered walls, width not exceeding 300, two coats 13 thick to blockwork base	M20.1.2.1
			&	
			Accessories, galvanised perforated metal angle bead, vertical, with 50 returns fixed to blockwork with masonry nails	M20.24.8 Angle beads taken to the vulnerable corners of the door opening.
			&	
			Two coats emulsion paint general plaster surfaces a.b. × 0.15 = m²	M60.1.0.1

External Doors 11.

0.94		Deduct Wrot Softwood skirting 19 x 75 a.b. & Deduct Knot, prime, stop, two undercoats one finishing coat gloss paint general isolated surfaces woodwork not exceeding 300 & Deduct Prime only ditto a.b. & Add Two coats emulsion paint general plaster Surfaces a.b. x 0.08 = m²	This demonstrates the use of 'Less' which reduces the length of both the 'deducts' and the 'add' above.
		[Reveals	
2/ 0.15		Less (x 0.08 = ____ m²)	Depending upon what was measured with the main structure, adjustment of floor, concrete bed and damp proof course to take.

14 Fixtures, sundries and stairs

This section of measurement includes such items as cupboards, wardrobes, kitchen units, hat and coat rails, notice boards, shelving, isolated handrails and balustrades and staircases. In certain contracts the work in this section could be quite extensive, for example a retail outlet may contain counters and storage facilities. Generally the work covered in this chapter is assumed to be in timber, other materials being considered beyond the scope of this volume.

Firstly it is necessary to establish if the items concerned are to be manufactured off–site and delivered to site in an assembled or a semi–assembled form. If this is the case, then the articles are classified as component items and their description is deemed to include breaking down for transport, installation and subsequent re–assembly. The description is invariably accompanied by a component detail, which shows sufficient information for the manufacture of the item, or a dimensioned diagram, which depicts the item being measured but probably leaves the form of construction to the discretion of the contractor. In all cases, as usual, a reference to a manufacturer's catalogue or standard specification could be used as an alternative.

Items which are made up on site such as isolated shelves, isolated worktops, unframed pinboards and hat and coat rails are measured as linear items and described with sizes which are deemed to be nominal unless described as finished. Ends, mitres and the like are deemed to be included. Labours which do not run the full length of the item are described as stopped and the number of them on each item is included in the description. Items such as backboards, plinth blocks, etc. are enumerated. Decoration of the items is taken separately.

Timber staircases are enumerated and may be either described fully with dimensions or accompanied by a component detail. Items such as linings, nosings, cover moulds, trims, soffit linings, spandrel panels, ironmongery, finishes, fixings, wedges and the like where supplied with or part of the component are deemed to be included. They would, of course, have to be included in the description or shown on the component detail. If these items are not part of the composite item then they are measured separately according to the appropriate rules. For example cover fillets and nosings are measured as linear items with ends being deemed included. The creation of the stairwell is usually measured with the floors section but it may be necessary to adjust the internal finishings at this stage. Decoration on the staircase itself, if not done at the factory, has to be measured. Whilst there are no special rules for painting on staircases remember that work in staircase areas has to be given separately. Handrails and balustrades which are isolated and do not form part of a staircase unit are measured as linear items and their size stated. Ramps, wreaths, bends, and ornamental ends on handrails are enumerated as extra over the handrail.

Fixtures and furniture such as counters, desks, benches, worktops, wardrobes, cupboards, shelving systems and kitchen units are enumerated and accompanied with a component drawing or dimensioned diagram. The drawing or diagram would show sufficient detail for the manufacture of the item and its fixing or placing in the work. As with staircases any adjustment to finishings or decorations, on–site decoration of the items and trim not supplied with the item have to be measured in accordance with the appropriate rules.

Taking-off List	**SMM Reference**
Fixtures, furnishings and equipment not associated with services	N10.1
Cover fillets etc. not included in above	P20.2
Ironmongery not included in above	P21.1
Adjustment of finishings and decoration	As before
Decoration on site	M60.1
Staircases	L30.1
Plus the items listed for fittings	As above
Isolated balustrades	L30.2
Ramps, wreaths, etc. to above	L30.4
Associated handrails	L30.3
Decoration	M60.1
Isolated shelves and worktops	P20.3
Unframed pinboards	P20.5
Backboards	P20.9
Decoration	M60.1

EXAMPLE 15

COMPONENT ITEM

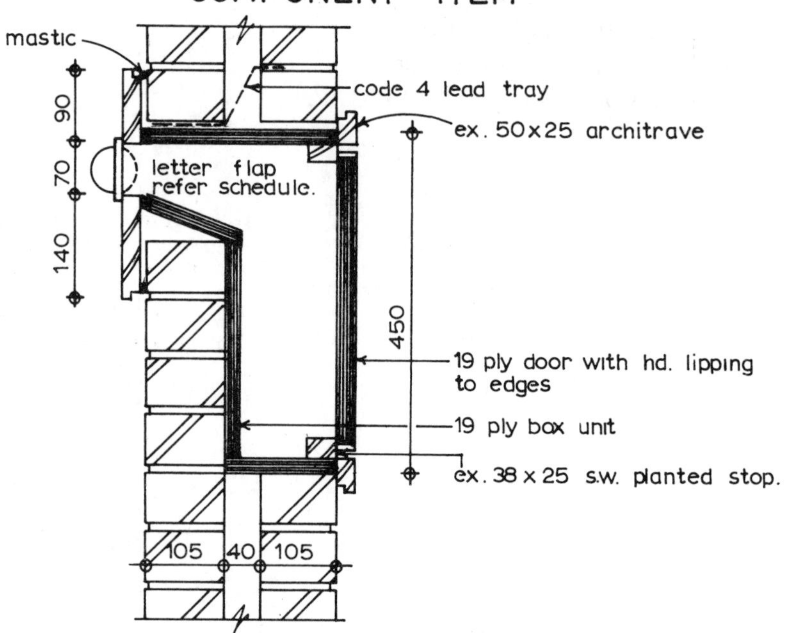

VERTICAL SECTION SCALE 1 : 100

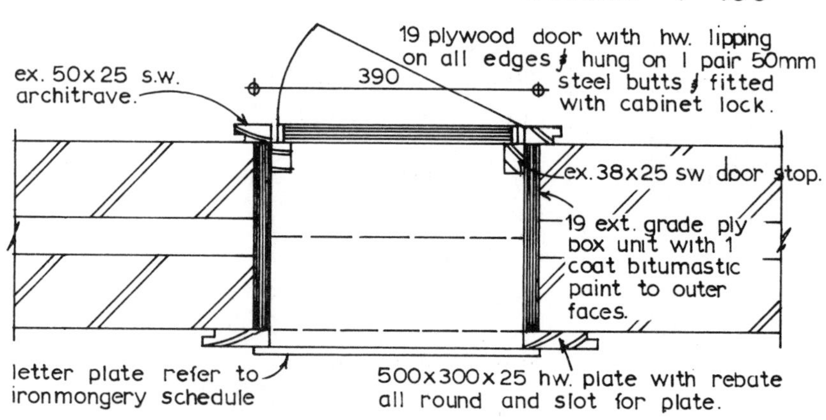

HORIZONTAL SECTION SCALE 1 : 100

NOTE: Aperture in ext. face brickwork to be 400x160.
Aperture in int brickwork to be 400x450.
To all timber k.p.s. & 3 coats gloss paint

FIGURE 35

Fittings 1.

COMPOSITE ITEM

2/105 · 210
40
2/25 = 50

300

1 | Letter box as component drawing Fig. 35, 500 x 300 x 500 comprising 19 plywood box lining treated one coat bitumastic paint to outer faces, 19 plywood inner door hardwood lipped on all edges, hung on & including pair 50 steel butts and fitted with cabinet lock & two keys, 38 x 25 wrot softwood door stops & 50 x 25 rebated architrave all round and 500 x 300 x 25 wrot hardwood external face plate rebated all round & slotted for letter (Contd.)

Being satisfied that the drawing complies with Rule S.1 of SMM N.10.1 the item is measured complete in the form in which in all probability it will be delivered to the site including in addition the fixing. The letter plate itself is excluded as this is part of the general ironmongery and would be fixed later and hence measured elsewhere.

Fittings 2.

			(Contd.) plate (measured separately) and fixing in prepared opening in cavity brick wall & pointing exposed edges in mastic	N 10.1.
			[Letter box.	
			105 100 30 235 2/150 = 390 300 690	
	1		Code 4 lead horizontal cavity tray 235 girth & 690 long and bedding in gauged mortar (1:1:6)	If the lead was measured superficial as required by F 30.2 the resulting quantity would be less than 1 m^2 and in accordance with GR3.3 would be given in the bill as 1 m^2. By enumerating, more meaningful information is given to the estimator. If there were many such cavity trays elsewhere on the project then this small quantity could have been added in.

		Fittings 3.		
1			Knot, prime, stop & paint two undercoats & one gloss finishing coat on isolated areas of woodwork not exceeding 0.50 m² [Inside	M60.1.0.3
			&	
			Ditto externally [faceplate.	There is no SMM requirement to measure the opening. The brickwork is not omitted and the only possible cost implication is temporarily bedding the bricks in sand courses until the item is ready for building in.

15 Plumbing

When measuring plumbing it is particularly important to follow a logical sequence of taking–off in order to be sure that no part is missed. Frequently, particularly on a small domestic installation, the only information shown on the drawings is the location of sanitary appliances. If this is the case then the measurement of the appliances is fairly straightforward and forms a logical start.

Having measured the sanitary appliances, and possibly coloured them in on the drawings, it is then easier to decide on a pipework layout for both wastes and supplies. Sizes of waste pipes are dictated by the size of the waste fitting from the appliance, e.g. wash basins 32 mm and baths and sinks 38 mm. Supply pipework sizes for a small installation should not be too difficult to assess. The rising main is usually in 15 mm pipework and the down feeds from the cistern in 28 or 22 mm reducing to 15 mm for the individual feeds except baths which require 22 mm. Adequate stopvalves, gatevalves and drain taps should be included in the system to enable sections to be isolated and drained.

Gatevalves do not restrict the flow of water when fully open and should be used on the low pressure distribution part of the system. Several water authorities require an indirect system with a drinking feed to the sink taken off the rising main and the remainder of cold feeds coming from a storage cistern which holds a reserve of water. The capacity of the cistern is given either as nominal, i.e. filled to the top edge, or actual, i.e. filled to the working water line. The requirements of water authorities vary considerably as regards required capacities but should be at least 112 litres actual for storage only, rising to 225 litres for storage and feeding a hot water system. The cistern may have to be raised to provide adequate pressure flow and the roof construction may have to be strengthened to support the additional weight. The inlet to the cistern is controlled by a float operated valve and the outlet should be opposite the inlet to avoid stagnation of water. An overflow pipe with twice the capacity of the inlet should be provided.

Before attempting to measure a plumbing installation trade catalogues depicting fittings available for the specified pipework should be obtained for reference. A selection can then be made of suitable fittings for connections to various appliances. A diagrammatic layout of the plumbing when provided is often not to scale and drawn in two dimensions. When measuring from such a diagram one has to visualise the layout in three dimensions and relate pipe runs to the structure. This will enable realistic lengths of pipes to be measured and the correct number of bends taken. Sometimes, when measuring copper pipes, it is difficult to decide whether to take 'made' bends (i.e. the pipe bent to form the bend) or fittings. Generally, made bends should only be taken for minor changes in direction of pipes or on short lengths. It should be remembered that long lengths of pipes with made bends may be impossible to install. For the measurement of installations without detailed information the following is suggested as a suitable order:

Sanitary applicances	(a)	sanitary appliances including taps, brackets, backboards, etc.
Foul drainage above ground	(b)	traps, wastes, overflows, soil and ventilating pipes including ducts
Cold water installation	(c)	connection to authority's main, supply to boundary of site and stop valve pit, reinstatement of highway
	(d)	supply in trench from boundary to building, stop valve, rising main to storage cistern
	(e)	branches from rising main
	(f)	storage cistern including bearers, overflow and insulation
	(g)	cold down services
Hot water installation	(h)	cold feed from cistern
	(j)	boiler and work in connection
	(k)	cylinder and main flow and return pipes
	(l)	secondary circulation, expansion pipe and branch services
Generally	(m)	casings
	(n)	testing etc.

Note: Insulation to pipes and builder's work in connection (e.g. chases, holes, painting etc.) should be taken after each sub–division.

An alternative approach to measurement is to follow the flow of water from the water main, through the building to the sanitary appliances and discharging into the drains. This is a more logical approach and would probably be adopted if the layout of the whole system is shown on the drawings. The main divisions shown above could still be used but in a different order.

As far as presentation in the bill is concerned, plumbing work has to be classified under headings indicating the nature of the work. For a simple domestic type installation these would be as follows:

(a) sanitary appliances
(b) foul drainage above ground
(c) cold water
(d) hot water
(e) sundry builder's work in connection with services
(f) testing and commissioning the drainage or water system

Note: For small scale installations (c) and (d) may be combined.

The builder's work in connection section may either be billed under a heading at the end of each appropriate work section or after the installation measurement.

When taking–off one obviously has to keep in mind these bill divisions but by following the suggested order of measurement given earlier the necessary sections will be automatically produced. There are several additional divisions of work for more sophisticated installations required by the SMM than those mentioned above but these are considered to be beyond the scope of this book.

SANITARY APPLIANCES

If sanitary appliances are specified fully then they are enumerated and the description should include the type, colour, size, capacity and method of fixing including details of supports, mountings and bedding and pointing. Frequently a catalogue or BS reference is used for part of the description but care must be taken to define any alternatives available. If full details are not available then a PC sum may be included for the supply of the appliances, an item included for contractor's profit and fixing measured as enumerated items. Descriptions should make it clear whether items such as taps and bath panels are included with the appliance. Small items such as towel rails and soap dishes must not be overlooked. Any builder's work such as tile splashbacks, bearers, backboards, painting and similar items necessary for the installation should be measured at this stage.

FOUL DRAINAGE ABOVE GROUND

Included in this section is the measurement of traps, waste pipes, overflow pipes and soil and ventilating pipes. Some appliances, such as water closets, have traps built in and some, such as wash basins, have integral overflows. Traps and other pipework ancillaries are enumerated with a dimensioned description and the method of jointing stated, although cutting pipes and jointing materials are deemed to be included. Bath and sink overflow outlets and flexible tubes to traps are also enumerated. Nowadays most waste and soil pipes are specified to be in plastic and their description should state whether they have ring seal or solvent welded joints and the type and spacing of pipe supports. Pipes are measured linear over fittings and joints in the running length, i.e. jointing straight lengths of pipe together, are deemed included. The nominal size of pipes has to be given and copper and plastic are usually described by their outside diameters and cast iron and mild steel by their nominal bores. Straight and curved pipes have to be classified separately and in the case of the latter the radius stated. Fixing the pipes to special backgrounds has to be given as follows:

(a) to timber including manufactured building boards
(b) to masonry which is deemed to include concrete, brick, block, and stone
(c) to metal
(d) to metal faced material
(e) to vulnerable materials which are deemed to include glass, marble, mosaic, tiled finishes or similar

Pipes laid in ducts, trenches, floor screeds and in situ concrete have to be so described. Fittings, such as bends and tees, to pipes not exceeding 65 mm diameter are enumerated and taken as extra over the largest pipe and are described as fittings with one, two or three ends stating if inspection doors are present. Fittings not falling within these categories are also measured extra over the largest pipe but are described with the method of jointing stated. Special joints and connections to different pipes and ancillaries are enumerated and described as extra over the pipe stating the method of jointing. Testing the foul drainage is given as an item, giving details of the tests and any attendance required. Cutting mortices and sinkings for the installation are enumerated stating their size and the nature of the structure and any necessary making good. Linear measurements are taken for cutting chases stating the number and sizes of pipes, the nature of the structure and any necessary making good. Holes for pipes enumerated and grouped as not exceeding 55, 55–110, and exceeding 110 mm nominal bore (SMM P31.20) and, although in some work sections holes are deemed included, those made at a later stage for services are considered to be measurable. Metal slates and collars and collars around pipes in asphalt and felt are enumerated. Painting pipes, described as painting services, is measured linear to pipes not exceeding 300 mm girth and superficial to those exceeding 300 mm girth. The measurement of overflow pipes to flushing cisterns must not be overlooked and these are measured in the same way as waste pipes.

COLD WATER

The measurement of the cold water installation will start invariably with the connection to the authority's main which would in all probability be included as a provisional sum. It is necessary to check with the authority the extent of work which they will carry out. Often included with the connection is the pipework to the stopvalve pit at the boundary and making good the highway. The stopvalve is either required to be located on the pavement or just inside the boundary. Pipes and fittings are measured in the same way as described for waste pipes. Stopvalves, gatevalves and ballvalves are defined as pipework ancillaries and are enumerated, described and the method of jointing stated. Storage cisterns are enumerated as general pipeline equipment and are described including the size and capacity. Overflows to cisterns should be taken at this stage.

Insulation to pipelines is measured linear and described including the thickness of the insulant and the nominal size of the pipe. Working insulation around ancillaries is enumerated as extra over the insulation. Insulation to equipment is either measured superficial (on the surface of the insulant) or enumerated giving the overall size. In the former case working around ancillaries is enumerated and in the latter case can be included in the item description. Excavating trenches for services not exceeding 200 mm nominal size is measured linear giving the average depth in 250 mm stages. Earthwork support, consolidation, backfilling and disposal are deemed to be included in the trench item. Stopvalve chambers and boxes are each enumerated and described. Underground ducts are measured as linear items giving the type, nominal size, method of jointing and whether straight or curved. Fittings and special treatment at ends are enumerated as extra over the ducts. Timber cistern bearers are measured linear and described as individual supports giving the cross section dimensions. The remainder of builder's work is measured as described before.

HOT WATER

Domestic hot water systems, apart from the pipework, have three main components, the boiler, the cylinder for storage and the cold feed storage. Suitable pipe sizes would be 28 mm for the primary flow and return between the cylinder and boiler and for the cold feed to the cylinder. The hot water distribution from the cylinder would be 28 mm reducing to 22 mm for the vent and to 15 mm for sink or basin supplies and 22 mm for the bath. These sizes should be regarded as minima and sizes would be increased for a larger number of draw off points. When an indirect heating circuit is included in the system then either a self–venting cylinder or a separate expansion and feed cistern have to be provided. Whilst heating installations are considered to be beyond the scope of this book it may be worth mentioning that a separate bill heading of low temperature hot water heating (small scale) would have to be introduced. Boilers and cylinders are enumerated and described under the rules for equipment and the description would have to include, as appropriate, the type, size, pattern, rated duty, capacity, loading and method of jointing. The remainder of the work is measured as described above.

Taking-off List	**SMM Reference**
Sanitary appliances	N13.4
Builder's work to ditto	P31 and appropriate work sections
Foul drainage above ground:	
Traps	R11.6.8
Waste pipes/soil pipes/overflow pipes	R11.1
Fittings to ditto	R11.2.3/4
Connections	R11.2.2
Builder's work in connection	P31
Cutting and forming holes	P31.20
Cutting mortices and sinkings	P31.21
Cutting chases	P31.22
Formwork to holes in concrete	E20.27
Holes in roof tiling	H60.11
Holes in metal roofing	H70–75.30
Metal slates	H70–75.26
Metal collars	H70–75.29
Backboards	P20.9
Asphalt collars	J21.18
Painting pipes	M60.9
Marking position of holes etc.	R11.12
Testing etc.	R.11.14
Cold Water Installation:	
Connection to main	Prov Sum
Stop valve	Y11.8
Main in trench	Y10.1.1.1.3
Fittings to ditto	Y10.2.3/4
Builder's work in connection	P30
Stop valve pit	P30.8
Stop valve box	P30.16
Trench for main	P30.1
Duct for main	P31.10
Fittings to ditto	P31.11.1
Stop valve and drain tap	Y11.8
Rising main and branches	Y10.1

Taking-off List (cont.)	**SMM Reference**
Fittings to ditto	Y10.2.3/4
Connections	Y10.2.2
Ballvalve	Y11.8
Storage cistern	Y21.1
Overflow to cistern	Y10.1
Fittings to ditto	Y10.2.4
Connections	Y10.2.2
Pipe insulation	Y50.1.1
Working ditto around ancillaries	Y50.2.1.1
Cistern insulation	Y50.1.4
Builder's work in connection	P31
Cistern bearers	G20.13
Remainder as before	See above
Pipework to down services	Y10.1
Fittings to ditto	Y10.2.3/4
Connections	Y10.2.2
Gatevalves	Y11.8
Insulation and builder's work as before	As above
Marking position of holes etc.	Y51.1
Testing etc.	Y51.4
Hot Water Installation:	
Boiler	T10(Y22.1)
Cylinder	Y23.1
Pipework, fittings, connections, insulation, testing and builder's work as before	

EXAMPLE 16

INTERNAL PLUMBING

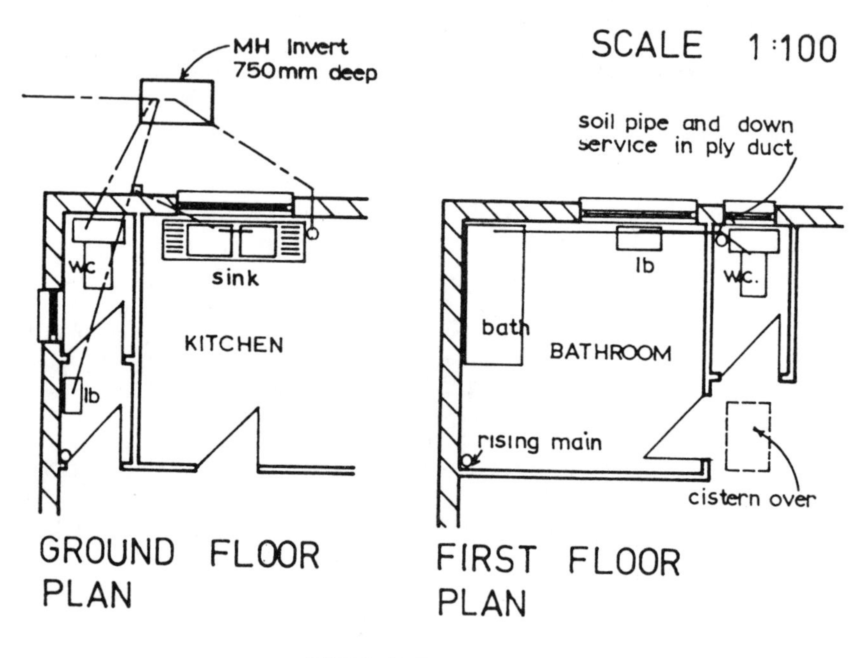

FIGURE 36

Internal Plumbing 1.

[Sanitary appliances.

P.C. Sum for Sanitary Appliances £ ______

A52
A p.c. sum is entered for the supply only on sanitary appliances assuming that insufficient detail is available to describe them fully.

Profit %

Fixing and assembling the following including bedding taps and waste fittings in mastic and bedding & pointing at abutments with walls in flexible sealant (joints to pipes measured)

2/ 1 White glazed vitreous china wash basins size 460 x 405 with pair chromium plated pillar taps, waste fitting, plug & chain and pedestal, including plugging & screwing brackets to masonry & bedding pedestal to floor & basin in mastic

N13.4.1

When describing the fixing of sanitary appliances it is desirable to give as much information as is known about the size and quality. The latter gives an indication of the amount of care required and replacement value if loss or damage should occur.

	Internal Plumbing 2.			
			[1st. Floor.	
	1		White glazed vitreous China W.C. suite with low level 9 litre plastic cistern and ballvalve, plastic flush pipe, seat & cover, including plugging & screwing cistern brackets to masonry & screwing pan to timber & bedding in mastic	
			[Grd. Floor.	
	1		Ditto but screwing pan to masonry & bedding in mastic	The ground floor w.c. is fixed to a solid floor.

Internal Plumbing 3.			
1		White reinforced acrylic bath size 1700 x 700 with pair chromium plated pillar taps, overflow fitting & flexible tube, waste fitting, plug & chain, moulded front and end panel including screwing adjustable bath feet to timber, plugging & screwing wall brackets to masonry and screwing panels to timber	

Internal Plumbing 4.

	1		Stainless steel double bowl, double drainer sink unit size 2000 x 600 with chromium plated mixer tap, overflow fitting and flexible tube, waste fitting, plug & chain, including setting in timber base unit (measured) and fixing with screws	N11.4.1 It is assumed that the timber base unit has been measured with other kitchen fittings.
7	1		Chromium plated toilet roll holders including plugging and screwing to masonry	Any other items such as mirrors, soap dishes and towel rails would be included here.

End of fixing & assembling

Internal Plumbing 5.

[Bath panel.

1700
700
2400

2.40 0.60			Sawn softwood framed supports width exceeding 300, 38 x 38 spaced at 600 c/cs	G.20.12.1 The framing for the bath panels together with any other ancillary work required for the sanitary appliances is measured here.

Foul drainage above ground

[Soil & vent pipe.

G. Floor. 2600
1st. " 2600
Floor. 250
5450

Roof space. 600
Above roof. 450
1050

5.45			Pipes, straight UPVC, 110 with ring seal joints & socket pipe clips plugged & screwed to masonry in duct	R11.1.1.1.2 Special mention has to be made of pipes fixed in ducts. Duct and holes for pipes assumed measured elsewhere.

Internal Plumbing 6.

	Qty		Description	Notes
	1.05		Ditto but not in duct	R11.1.1 The soil pipe in the roof space. The branch from the first floor W.C. and the pipe from the ground floor W.C. to the drain is added.
	0.80		[1st. flr. W.C.	
	0.30		[Gd. flr. W.C.	
	1		Extra over ditto for UPVC vent terminal solvent welded	The vent in the top of the soil stack.
2/	1		Extra over 110 UPVC pipe for joint to 100 clayware drain pipe socket including UPVC drain adaptor and gaskin & Cement & Sand (1:2) joint	R11.2.2 Joint at base of soil pipe to drain, and joint of pipe from ground floor W.C. to drain.
2/	1		Ditto for bent W.C. Connector & Ditto for joint of W.C. outgo to socket including W.C. adaptor size 108 – 114	Special W.C. bend with enlarged socket.

Internal Plumbing 7.

Timesing	Dimension	Squaring	Description	Side notes
	1		Ditto for 110 boss branch including ring seal joint	Branch on soil pipe for 1st. floor W.C. branch connection.
			[Traps.	
			[Wash basins.	
2/	1		36 Polypropylene tubular swivel 'P' trap with 76 seal including screwed mastic joint to waste fitting and ring seal joint to MUPVC pipe	R11.6.8.1
			[Sink	
	1		42 Ditto	
			[Bath.	
	1		42 Ditto bath trap with 36 overflow bend including do.	

Internal Plumbing 8.

			Description	Notes
			[Wash basins.	
	0.60		Pipes, straight MUPVC 36 with solvent welded joints and clips plugged & screwed to masonry [GF.	R11.1.1.1
	1.50		[1st. flr.	
	3.50		Ditto 42 do. [Bath.	
	0.80		[Sink.	
			[Wash basins.	
2/	2		Extra over 36 MUPVC pipe for fittings two ends	R11.2.3.2 Fittings to pipes not exceeding 65 diameter are described by the number of ends. These dimensions represent the bends on the waste pipes.
			[Bath & Sink.	
2/2/	1		Ditto 42 do.	
			[1st. flr. W/B waste.	
	1		Extra over 110 UPVC pipe for 36 boss branch with ring seal joint	R11.2.4 The boss on the main soil stack for the first floor wash basin waste.

			Internal Plumbing 9.	
			[Bath waste.	
	1		Ditto 42 boss branch do.	Ditto bath waste.
			[G.F. w/B waste.	
	1		Extra over 36 MUPVC pipe for joint to 50 cast iron drain pipe socket with UPVC socket reducer & caulking bush including gaskin & cement & sand (1:2) joint	The joint of the ground floor wash basin waste to the drain.
			[Sink.	
	1		Ditto 42 MUPVC pipe do.	Ditto sink.

Internal Plumbing 10.

Timesing	Dimensions	Squaring	Description	Ref.
			Overflows to W.C. cisterns.	
7/	1.00		Pipes, straight UPVC 22 overflow with solvent welded joints & clips plugged & screwed to masonry	
7/	2		Extra over ditto for fittings two ends & Ditto for connection to flushing cistern including straight tank connector	
	Item		Marking the positions of holes, mortices and chases in the structure for the foul drainage above ground	R11.12 .1
	Item		Testing & commissioning the foul drainage above ground installation	R11 14.1

Internal Plumbing 11.

Builders Work in connection with Plumbing

P30/31. M2

[Soil & Vent pipe.

1			Cutting holes in roof tiling for pipe 55 – 110 diameter	H60.11
			&	
			Aluminium patent weathering slate 457 x 400 with moulded rubber sealing cone for 110 UPVC pipe handed to others for fixing	H72.26.1.1.1
			&	
			Fixing only ditto (by roofer)	H60.10.0.0.1
1			Cutting hole in 25 softwood board flooring for pipe 55 – 110 diameter	K20.10. P31.20.2.2

Internal Plumbing 12.

Builders Work (Contd.)

1st. Flr. W/B & bath waste.

2/ 1	Cutting hole in 100 block wall for pipe not exceeding 55 diameter	P31.20.2.1

End of Builders Work.

Piped Supply Systems

Water Supply (Cold Water)

Y10 (M1)

Copper tubes to be to BS 2871 Part 1, half hard, table X

It is assumed that the connection to the main, stopvalve pit and incoming main to the stopvalve in the building has been measured elsewhere.

Fittings for copper tubes to be to BS 864 Part 2 compression type A (non manipulative)

Insertion of preamble items common to several descriptions saves repetition.

Copper tube to be fixed with standard saddle band type clips spaced at 1200 c/cs fixed with brass screws.

Internal Plumbing 13.

Timesing	Dimension	Squared	Description	Notes
			G. Floor. 2600 1st. " 2600 Floor. 250 5450	
	5.45		Pipes, straight copper 15 clips plugged & screwed to masonry	Y10.1.1.1 The rising main and the branch to the sink for drinking water.
	6.40		[To sink 3200 2500 700 6400	
			4000 Clg. 100 Vert. to cist. 500 4600	
	4.60		Ditto but clips screwed to timber	The rising main in the roof.
			[Roof.	
4/	1		Extra over ditto for fittings two ends	Y10.2.3.3 Bends on pipes.
2/	1		[Sink.	
			[Sink branch.	
	1		Ditto three ends	Tee for sink branch.

Internal Plumbing 14.

			[Ball valve Sink tap.	
2/	1		Extra over ditto for joint to fitting including straight connector	Y10.2.2.1 Joints to threaded end of ball valve and sink tap.
			[Stopvalve at entry.	
	1		Combined high pressure screw down stop valve and drain tap as BS 1010 & 2879 A, compression joints to copper tube	Y11.8.1.1 The stopvalve and drain tap on the rising main at the point of entry to the building.
			[Cistern in roof.	
	1		Water storage cistern to BS 4213 polythene ref. PC 60; 273 litres capacity, including lid, perforations for one 15 three 22 pipes and backing plates (connections measured) & placing in position	Y21.1.1.1 The water storage cistern in the roof space.

Internal Plumbing 15.

	1		Ballvalve, high pressure; BS 1968 Class A PVC float 15 inlet fixed to polythene cistern and connection to copper pipe including straight coupling	Y11.8.1.1 The ballvalve controlling the flow of water into the cistern.
			[Overflow.	
	3.50		Pipes, straight UPVC 22 overflow with solvent welded joints & clips screwed to timber	The overflow to the storage cistern.
3/	1		Extra over ditto for fittings two ends	Bends on the overflow pipe.
	1		Ditto for connection to polythene cistern including straight tank connector	The connection of the overflow to the cistern.

Internal Plumbing 16.

			Description	Reference
	Item		Marking the position of holes, mortices & chases in the structure for the cold water installation	Y51.1
	Item		Testing and Commissioning the cold water installation	Y51.4.
			Builders Work in Connection with plumbing	
			[Rising main.	
	1		Cutting hole in 25 softwood board flooring for pipe not exceeding 55 diameter	P31.20.2.1 NOTE: Holes in plasterboard and making good plaster around pipes not considered measurable.
			[Sink branch.	
2/	1		Cutting hole in 100 block wall for pipe not exceeding 55 diameter	P31.20.2.1

Internal Plumbing 17.

Times	Dimensions	Squaring	Description	Reference
	5.45 6.40		Prime & two coats gloss paint copper services, isolated surfaces not exceeding 300 girth	M60.9.0.2 Painting on exposed pipes.
			[Cistern bearers.	
3/	1.50		Individual supports 50 × 100 sawn softwood pressure impregnated	G20.13.0.1
			[Cistern platform.	
	1		Backboards etc. 1500 × 1250 chipboard 19 BS 5669 flooring grade	P20.9.1
			[Overflow thro' fascia	
	1		Cutting hole in 25 softwood fascia for pipe not exceeding 55 diameter	

End of Builders Work

			Internal Plumbing 18.	
			Insulation to cold water supply	Y50 (M1)
			[Pipes in roof.	
4.60			Insulation 20 thick glass fibre sectional to 15 diameter copper pipes including metal bands	Y50.1.1.1 Working around fittings is deemed included but but working around ancillaries is enumerated.
			[Cistern.	
1			Insulation 25 thick expanded polystyrene boards to sides & top cistern 273 litre capacity including securing with twined steel bands and cutting around pipes	Y50.1.4.2 Insulation to the cistern.
			NOTE. Down services and builders work in connection to measure.	NOTE: The down services have been omitted from this example as the work is largely a repetition of what has already been measured.

16 Drainage

The measurement of drainage may be divided into the following sections:

(1) manholes or inspection chambers
(2) main drain runs and fittings between manholes
(3) branch drain runs and fittings between outlets and manholes
(4) accessories (e.g. gullies)
(5) sewer connection
(6) land drains
(7) testing

If the foul water and the surface water drainage are separate systems then usually each would be measured independently using the above order. There may, however, be a possibility of combining manhole measurement if the construction is similar. When measuring surface water drainage there must be liaison with the person measuring the roads and pavings to ascertain the position of gullies and channels. Similarly there will have to be consultation with the person measuring the roofs to ascertain the position of rainwater pipes and to decide who will measure the connections if required. When measuring foul water drainage there will have to be similar discussion with the person measuring the plumbing to find out the position of soil pipes and outlets and again to establish who is to measure the connections. The opportunity should be taken during the discussions to see that the drawings show drain runs leading from all service outlets.

In most cases the position of manholes will be shown on the drawings together with invert levels. If the existing and finished ground levels adjacent to the manhole are not shown then these will have to be ascertained from the site plans, possibly by interpolation. A basic decision will have to be made as to whether excavation is to be measured from the existing ground level or, if there is excavation for other work in the vicinity of the manhole, from the reduced level. Usually the reduced level excavation takes place prior to the drain excavation and it would be correct to excavate drains from the reduced level, although one should state in the bill the method used. If there is filling adjacent to the manhole then one has to assess whether the filling is likely to be carried out prior to the drain excavation. When this is the case the depths could be taken from the cover level as excavation through fill is measured as normal excavation. If the filling is likely to be carried out after the drain excavation, for example hardcore fill under roads, then the drain excavation would be measured from the original ground level.

Prior to the measurement of manholes it is wise to prepare a schedule of similar format to that shown in Appendix 2. The schedule will enable manholes with similar plan dimensions to be spotted. Manholes with the same plan dimensions but of differing depths may have their depths averaged for measurement purposes, although for excavation they must be within the same depth category. If information on the sizing of manholes is not available then the following guide may be useful:

- For inspection chambers up to 900 mm deep minimum internal size 700 mm wide × 750 mm long, allowing up to two branches on both sides.
- For manholes over 900 mm and up to 3300 mm deep, the minimum size should be 750 mm wide × 1200 mm long, allowing 300 mm in the length for each 100 mm branch and 375 for each 150 mm branch.
- For manholes up to 2700 mm deep a cover size 600 × 600 mm should be provided, increasing to 600 × 900 for deeper manholes.
- For deep manholes an access shaft can be constructed at the top to within 2 m of the benching.
- When the depth of the manhole exceeds 900 mm step irons should be taken at 300 mm intervals.

The measurement of manholes generally is carried out as for other building work. Excavation is described as pits and the number stated in the description. Precast concrete cover slabs are enumerated but in situ slabs will have to be measured in detail with separate items for formwork and reinforcement. Building in ends of pipes, channels, benching, step irons, covers and intercepting traps are enumerated and described giving sizes. If pre-formed manholes are used then the excavation and in situ concrete work in association is measured in the same way but the unit itself is enumerated and fully described.

Once the manholes have been measured it is a comparatively simple matter to measure the main drain runs between. Firstly a schedule should be prepared on the lines of that shown in Appendix 2. Depths of the drain excavation will usually be the same as that for the manhole at each end, although a small adjustment may have to be made if the bed below the manhole is of different thickness to that below the drain pipe. The depths at each end of the drain run will be averaged and the depth classified in 250 mm stages. Trenches for pipes are measured as linear items and those up to 200 mm diameter can be grouped together and for those over 200 mm the size is given. Earthwork support, treating bottoms, filling and disposal are all deemed to be included with the trench item.

Bedding for pipes is classified as one of the following:

(a) beds
(b) beds and haunchings
(c) beds and surrounds
(d) vertical casings

In the case of (b), (c) and (d) the size of the pipe has to be given. In the case of (a) and (b) the width and thickness of the bed have to be given and in (c) the thickness of the surround in addition. In the case of (d) the size is given.

Pipes are measured as linear items measured over all pipe fittings, described as in trenches with the nominal size and method of jointing. Strictly speaking the pipe length will be longer than the trench to allow for building in to the sides of the manhole. In practice, however, it is unlikely that measurement could be that accurate, particularly if the drawing is to a small scale. Pipe fittings such as bends are enumerated as extra over the pipe.

Branch drain runs are measured in the same way as main drain runs and again a schedule should be prepared prior to measurement. The important difference is that branch runs are connected to the manhole at one end only. At the other end will either be a gully or a connection to a soil pipe or similar outlet. The depth of the trench at this end will depend upon the size of the gully or rest bend but for normal circumstances could be taken as 600 mm. If bends are used then remember that the pipe must be measured to include their length. Gullies and other accessories are enumerated and described, jointing to pipes and concrete bedding being deemed included. Back and side inlets, raising pieces and gratings should be included in the description of the gully. Usually two bends are required in a pipe coming from a gully as there may have to be a change in direction as well as an adjustment to achieve the correct fall. Circular top or two piece gullies may, however, reduce the need for direction change.

The connection to the publicly owned sewer, if carried out by the contractor is enumerated and described. If the connection is carried out by the statutory authority, as is usually the case, then a provisional sum is included to cover the cost of the work. Care must be taken to ascertain the extent of the work done by the authority and to include for the remainder in the measured work. The work frequently involves excavation across the highway which would require special provisions such as traffic control, lighting and reinstatement. It should also be noted that work beyond the boundary of the site has to be kept separately.

Testing the drainage system is included as an item stating the method to be used. The extent of testing required should be ascertained from the appropriate authority.

Taking-off List	**SMM Reference**
Manholes:	
Excavation	D20.2.4
Removal of surplus	D20.8.3
Earthwork support	D20.7
Compacting bottom	D20.13.2.3
Concrete base	E10.4
Brick sides – fair face	F10.1
Rendering	M20.1
In situ concrete cover slab	E10.5
Reinforcement	E30.1–4
Formwork to soffit	E20.8
Formwork to edges	E20.3
Formwork to hole	E20.27
Precast concrete slab	F31.1
Cover and frame	R12.11.11
Lifting keys	R12.11.13
Benching	R12.11.9
Main channel	R12.11.8
Branch bends	R12.11.13
Building in ends of pipes	R12.11.7
Step irons	R12.11.10
Intercepting traps	R12.11.12
Backfilling	D20.9
Preformed manholes	R12.11.14
Main drain runs:	
Excavating trenches	R12.1
Beds, haunchings and surrounds	R12.4/5/6
Pipes	R12.8
Pipe fittings	R12.9
Branch drain runs:	
Excavation, beds, pipes and fittings	as above
Accessories (gullies etc.)	R12.10
Work outside the boundary of the site:	Gen 7.1(c)
Connection to sewer	R12.16
Disposal of ground and surface water	R12.3
Testing	R12.17

EXAMPLE 17

DRAINAGE

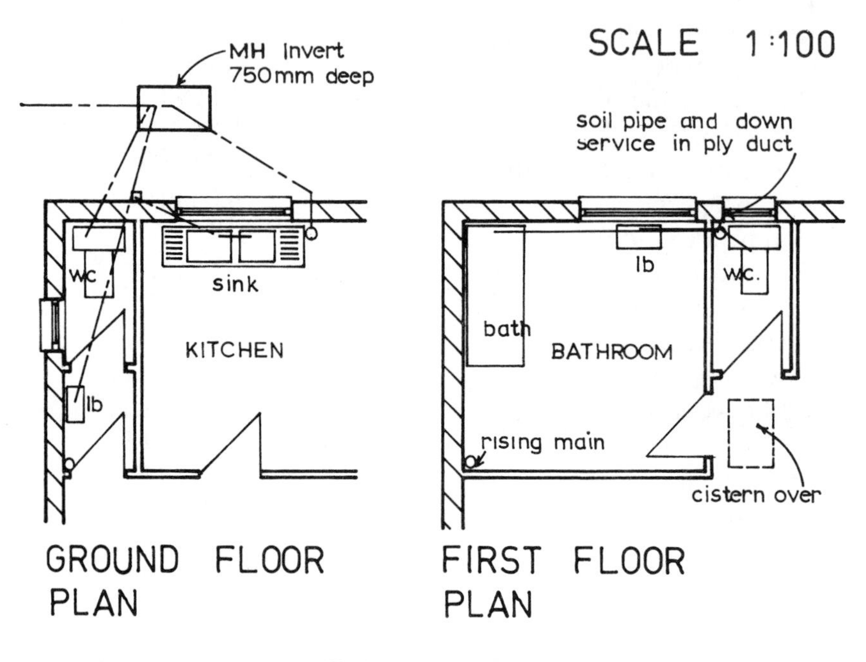

FIGURE 37

Drainage 1.

Nr. 1 Manhole.

900 x 570 x 750
to invert -
no spread.

	900	x	570
2/215 =	430		430
	1330		1000

To invert.	750
Bedding.	25
Bottom.	150
	925

The SMM does not require manholes to be kept under a heading but this is considered desirable for estimating purposes.

1.33 1.00 0.93	Excavating pit maximum depth not exceeding 1.00 m & Disposal of excavated material off site	R12.12.1 R12.12.1

The number of pits is indicated in the heading and need not therefore be repeated in the description.

	1330
	1000
2/	2330
	4660

Drainage 2.

Nr. 1 Manhole (Contd.)

Dimensions		Description	Ref.
4.66 0.93		Earthwork support maximum depth not exceeding 1.00 m, distance between opposing faces not exceeding 2.00 m	R12.12.1
1.33 1.00		Surface treatment compacting bottoms of excavations	R12.12.1
1.33 1.00 0.15		In-situ concrete (20 N/mm²) bed not exceeding 150 thick poured on earth	R12.12.2
		900 570 2/1470 = 2940 4/215 = 860 3800	
		To invert. 750 Bedding. 25 775 Cover slab - 100 675	

Drainage 3.

Nr. 1 Manhole (Contd.)

Dim.		Description	Ref.
3.80 0.68		One brick wall, vertical, in semi-engineering bricks, English bond in cement mortar (1:3)	R12.12.5
1		Bending 900 x 570 average 190 thick in concrete (20 N/mm²) finished in cement & sand (1:2) trowelled smooth to steep falls & channels	R12.12.9
		3800 4/215 = 860 2940 675 - 190 485	
2.94 0.49		Rendering to walls exceeding 300 wide in cement & sand (1:2) 19 thick, one coat trowelled smooth to masonry	R12.12.6

Drainage 4.

Nr. 1	Manhole (Contd.)		
1		100 Vitrified clayware half round curved main channel 1100 girth & bedding in cement mortar (1:3)	R12.12.8
2		100 Ditto three quarter section branch channel bend ditto	R12.12.8
4		Building in end of 100 pipe to one brick wall	Cutting pipes deemed included (C6). R12.12.7
1		Precast Concrete (25 N/mm²) cover slab 1000 x 700 x 100 thick reinforced with steel fabric to BS 4483 ref. A142 with rebated opening 610 x 475, finished smooth on top & setting in cement mortar (1:3)	R12.12.11

Drainage 5.

Nr. 1 Manhole (Contd.)

1			Galvanised mild steel single seal solid cover & frame 610 x 457 & setting frame in cement and cover in grease	R12.12.11
			[No step irons.	

(End of Nr. 1 Manhole.)

			[Branches into MH.	The average depth of the trench is arrived at by calculating the average between the MH invert level of 750 and an assumed level leaving the fittings.
			750 450 2) 1200 av. 600 Conc. 100 700	
1.60			Excavating trenches for pipes not exceeding 200 diameter, average depth 500 - 750 [WC.	If special materials are required for back filling details would be stated.
1.00			[Gully.	
2.10			[SP.	R12.1.1.2

Timesing	Dimensions	Squaring	Description	Notes
			Drainage 6.	
	2.20 1.00 2.70		100 Vitrified clayware pipe with flexible mechanical spigot & socket joints in trenches	600 mm added for wall thickness in case of W.C. and soil pipe from above. R12.8.1.1
	1.60 1.00 2.10		In-situ concrete ($20 N/mm^2$) bed & haunching to 100 pipe, 425 wide & 100 thick	R12.5.1.1
			100 = 125 2/150 = 300 425	Outside diameter equals 125 mm plus spread of 150 mm each side.
3/	2		Extra over 100 pipe for bend	2 bends assumed for each run at fitting. R12.9.1.1
	1		Clayware trapped gulley with 100 outlet, 225 raising piece 225 high with two 100 back & side inlets, 100 x 50 reducing sockets set in inlets, galvanised grating	R12.10.1.1

	Drainage 7.		
		[Branches into gulley.	
1.40 2.20		Excavating trenches for pipes not exceeding 200 diameter, average depth not exceeding 250 [Sink. [G.F. basin.	
1.50 2.30		50 Cast iron pipe with spigot & socket joints in tarred gaskin & molten lead, in trenches in runs not exceeding 3m long (In Nr. 2)	R12.8.1.1.1
1.40 2.20		In-situ Concrete (20 N/mm²) bed & surround to 50 pipe 375 wide & 275 thick	The jointing of the cast iron pipes to the gully inlets is deemed included. R12 (C5) R12.6.1.1
2/ 2		Extra over 50 pipe for bend	2 bends per run as before.

Drainage 8.

Testing - as standard.

R12.17

Alternatively if the manhole was of preformed construction it would have been enumerated as R12.11.14 stating all the details as set out in the rule.

To TAKE -

Main drain Vent.

Connection to Sewer.

17 External works

Before the measurement of external works is commenced a visit to the site should be made to ascertain items to be included or to check the information shown on the drawings. Items which could be checked or taken include:

(a) grid of levels and dimensions of the site
(b) pavings etc. to be broken up
(c) demolition of walls, fences, buildings etc.
(d) felling trees and grubbing up hedges
(e) preservation of trees and grassed areas etc.
(f) any existing services, overhead powerlines etc.
(g) any turf which may be worth preserving

In addition to noting these items consideration could be given at the same time to other matters, such as access to the site, which may have to be drawn to the attention of the tenderers in the preliminaries.

The measurement of external works can include several different aspects of work and to give an idea of the coverage a selection of items is listed below.

(1) Site clearance including:
 (a) removal of trees, hedges and undergrowth
 (b) lifting turf for re-use
 (c) breaking up pavings etc.
 (d) demolition work

(2) Temporary works (other than those for the contractor's own convenience) including:
 (a) roads
 (b) fencing
 (c) maintaining existing roads

(3) Roads, car parks, paths and paved areas including:
   (a) preparatory work
   (b) pavings
   (c) curbs, edgings, channels
   (d) drop curbs
   (e) steps and ramps
   (f) road markings
   (g) surface drainage

(4) Fencing and walls including:
   (a) boundary, screen, retaining walls
   (b) fencing, gates
   (c) guard rails
   (d) in situ planters

(5) Outbuildings including:
   (a) garages
   (b) sub-stations
   (c) gate-keepers offices
   (d) bus shelters
   (e) canopies

(6) Sundry furniture including:
   (a) bollards
   (b) seats and tables
   (c) litter, grit and refuse bins
   (d) cycle stands
   (e) prefabricated planters
   (f) flag poles
   (g) clothes driers
   (h) sculptures
   (j) signs, notices
   (k) lighting standards

(7) Water features including:
   (a) lakes, ponds
   (b) ornamental and swimming pools
   (c) fountains

(8) Horticultural work including:
   (a) cultivating, topsoil filling, sub-soil drainage
   (b) grassed areas (seeding and turfing)
   (c) planting trees (tree grids and guards)
   (d) planting shrubs, hedges and herbaceous plants

(9) Sports facilities including:
   (a) playing fields, running tracks
   (b) tennis courts
   (c) bowling greens
   (d) children's play equipment

(10) maintenance including:
   (a) planted and grassed areas
   (b) playing fields

(11) External services including:
   (a) water, gas and electric mains
   (b) telephone and TV services
   (c) fire and heating mains
   (d) security systems

Whilst the above areas of work have been classified under various headings the presentation of the work in the bill is a matter for personal preference. The SMM gives work sections for land drainage, roads and pavings, edgings, site planting, fencing and site furniture and it would be as well to follow these at least. As far as services are concerned, pipes in trenches have to be kept separately. Special rules apply to the measurement of trenches for services which are measured as linear items classifying as for pipes not exceeding 200 mm nominal size and if over 200 mm, stating the size. Average depths are stated in 250 mm stages and earthwork support, backfilling and disposal are deemed included.

Removing trees and tree stumps from the site are taken as enumerated items, giving the girth as 600 mm to 1.50 m, 1.5 to 3 m and exceeding 3 m. Grubbing up roots, disposal and filling voids are deemed to be included although a description of the filling has to be given. Lifting turf for preservation is given as a superficial measurement. General excavation in connection with external works is measured under the same rules as for general building work. Walling in connection with external works is again measured in accordance with the general rules although one is more likely to come across curved and battered work.

The main problems in the measurement of external works are likely to arise from excavating to reduce levels in uneven ground and measuring irregular areas. Guidance on calculations for these items is given in Chapter 4. Provided that a proper grid of levels has been taken over the site it should be fairly simple to find the average depth of excavation.

There may be a difficulty in giving the required depth stages for excavation when depths vary over the site. If this appears to be a problem and would mean drawing many contours representing depth changes, it is suggested that, irrespective of the required depth stages, an average depth is found for the area and this is used for the classification. The contractor should, of course, be informed that this method has been adopted. Frequently it is prudent to divide large irregular areas into sections for measurement; not only will this assist with the calculation of the area but also give a greater accuracy with the depth measurements and classifications. Furthermore it may be wise to make a division at the edge of an area of shallow excavation where it becomes deeper. Regarding earthwork support it should be remembered that this is required to be measured only if the excavation exceeds 0.25 m in depth. Handling excavated material generally is the responsibility of the contractor but if there are any specified conditions regarding handling, such as the provision of temporary spoil heaps, then these have to be stated.

Specifications for roadworks frequently refer to guides and recommendations published by the Department of Transport, the Department of Environment or the Transport and Road Research Laboratory. Concrete roads are measured as cubic items and classified as for slabs. Mechanical treatment to the surface of the concrete is given as a superficial item. Macadam roads are measured superficial, the area taken being that in contact with the base. No deductions are made for voids within the area when not exceeding 0.50 m^2. Work is described as being as level and to falls, to falls and crossfalls and slopes not exceeding 15 degrees slope or to slopes exceeding 15 degrees. Forming or working into shallow channels is included but linings to large channels are taken as linear items stating the girth on face; all labours being deemed included. Gravel paving or roads are measured and classified in the same way as macadam.

Brick and block paving is measured in a similar manner and classification to gravel, except that measurements are taken on the exposed face. The SMM includes rules for the measurement of special pavings for sport, these are measured the area in contact with the base and are again classified under the same categories mentioned above. Kerbs, edgings and channels are measured linear with 'specials' being enumerated as extra over items.

Fencing is measured as a linear item over the posts. There are several specifications for fences and these can frequently be used to assist with descriptions. Posts or supports occurring at regular intervals are included in the description of the fencing but occasional supports such as straining posts are enumerated and described as extra over the fencing. The excavation of post holes, backfilling, earthwork support and disposal of surplus are deemed included but the size and nature of the backfilling has to be stated. Gates are enumerated and ironmongery is also enumerated separately.

Most of the items listed above under the heading of sundry furniture are enumerated and supported by a component drawing, dimensioned diagram or reference to a trade catalogue or standard specification.

The taking–off list which follows can only, of course, show the more important sections involved in this wide area of work and it is suggested that this is used in conjunction with the coverage list given earlier in this chapter.

Taking–off List	**SMM Reference**
Site preparation	D20.1
Excavation, earthwork support, filling and surface treatment	See substructure
Breaking out existing materials	D20.4/5
Concrete roads and pavings	Q21.1–5
Macadam pavings and roads	Q22.1/2
Gravel roads and pavings	Q23.1/2
Interlocking block roads and pavings	Q24.1/2
Block roads and pavings	Q25.1/2
Kerbs, edgings and channels	Q10.2–5
Special surfacings for sport	Q26.1–9
Seeding and turfing	Q30.1–6
Planting	Q31
Fencing	Q40
Site furniture	Q50.1
Land drainage	R13
Trenches for engineering services	P30
Holes, chases etc., for services	P31

EXAMPLE 18

ROADS AND PATHS

Roads and Paths 1.

(Length of road assumed 45 m)

		Road.
Road	3600	254
Kerb. 2/127 =	254	150
Spread. 2/162 =	324	404
	4178	

ROAD WIDTH 3600

127 × 254 KERB

150 Concrete road

150 Hardcore

150

450

Timesing	Dimensions	Description	Notes
	45.00 4.18 0.40	Excavating to reduce levels not exceeding 1.00m deep	D20.2.2
		&	
		Disposing excavated material off site	D20.8.3.1
	45.00 4.18	Compacting bottoms of excavation	D20.13.2.3 The SMM does not require the measurement of compacting to include the fact that the work is to falls and or cambers.
2/	45.00 0.40	Earthwork support maximum depth not exceeding 1.00m, distance between opposing faces exceeding 4.00m	D20.7.1.3

Roads and Paths 2.

	- 2/162 = 3600 324 3276	
45.00 3.28 0.15	Filling making up levels average thickness not exceeding 0.25m with hardcore obtained off site compacted by vibrating roller	The words 'obtained off site' would probably be included in a heading to save repeating everytime a hardcore item arises. Q 20.10.1.3
45.00 3.28	Surface treatment, compacting and blinding hardcore filling to falls and cambers and blinding with ashes & Damp proof membrane width exceeding 300, horizontal, waterproof building paper to BS 1521 grade B1 lapped 225 at joints and laid on blinded hardcore bed to receive concrete road	Q 20.13.2.2.1 Because the hardcore filling has been measured cube, work to the top surface is required to be measured separately and described. Compacting to falls and cambers of a hardcore bed is cost significant and therefore warrants mentioning.

Roads and Paths 3.

Timesing	Dimensions	Squaring	Description	Notes
	45.00 3.60 0.15		In-situ concrete ($20 N/mm^2$) bed thickness not exceeding 150 in bays not exceeding $40 m^2$ each including formwork between bays	Q 21.1
	45.00 3.60		Reinforcement for in-situ concrete, steel fabric reinforcement to BS 4483 Ref. B 785 lapped 150 at joints	If for any reason there was a bent edge to the fabric it would be clearer to an estimator to measure a linear item of 'extra over' rather than describe the whole area as 'bent'. Q 21.3
4/	3.60		Designed joint in in-situ concrete, depth not exceeding 150, with pre-moulded impregnated fibre board, the top 15 filled with approved sealing compound to BS 2499 and cutting fabric	Assumed every 10.00 m Q 21.4

Roads and Paths 4.

Timesing	Dimensions	Squaring	Description	Notes
2/	45.00		Precast concrete (granite aggregate) kerb to BS 340 Fig. 5, 127 × 254, bedded and jointed in cement mortar and haunched in concrete (10 N/mm²), including 450 × 150 concrete (10 N/mm²) kerb foundation	Cut angles and ends are included (C1) but standard items such as quadrant ends, dropped kerbs etc. would be specials under Q 10.5 Q 10.2.1.0.2
2/	45.00 0.16 0.25		Deduct Disposing excavated material off site & Add Filling to excavations average thickness not exceeding 0.25 m with excavated material arising from the excavations, compacted in maximum 100 layers	A small adjustment required to substitute filling for the volume behind the kerb already removed from site. Q 20.9.1.1

Roads and Paths 5.

Path.

12.00 1.20		Excavating topsoil for preservation average 150 deep	This time the top of the excavation is assumed to be in topsoil to be retained. D20.2.1.1
		&	
		Disposal of excavated material on site, deposit in heaps a distance not exceeding 100.00m	D20.8.3.2
		× 0.15 = m³	
		&	
		Compacting bottoms of excavations	D20.13.2.3
12.00 1.00 0.10		Filling to excavations average thickness not exceeding 0.25m with hardcore obtained off site	Q20.9.1.3

	Roads and Paths 6.			
	12.00 1.00		Compacting surface of hardcore filling, blinding with fine ashes	Q 20.13.2.2.1
			&	
			Paving, two coat tar to BS 1242 level, to falls and crossfalls, base course 30 thick, wearing course 20 thick rolled and blinded with grit	Q 22.2.1
	12.00		Precast concrete (granite aggregate) path edging to BS 340 Fig. 10 51 × 152, bedded and jointed in cement mortar, including 150 × 75 concrete (10N/mm²) edging foundation poured against earth	Assumed one side only. The other made good to road kerb. Q 10.2.1

18 Demolitions, alterations and renovations

As the work covered by this chapter relates entirely to existing buildings it should be noted that all work on, or in, or immediately underneath work existing prior to the current project must be kept in a separate section of the bill.

Traditionally alteration works have been referred to as 'spot items', 'works on site' or 'works to be priced on site' which incorporated demolition, repair and alteration work. The SMM now subdivides the work as follows:

(a) demolition
(b) alterations – spot items
(c) repairing/renovating concrete/brick/block/stone
(d) Chemical damp proof courses to existing walls
(e) repairing/renovating metal and timber
(f) fungus/beetle eradication

In addition to the above the SMM includes a section of additional rules for work to existing buildings covering work groups H, J, K, L, M, R and Y.

The main points to remember when measuring this type of work are firstly to identify carefully the exact location of the work, secondly to describe clearly what has to be done and thirdly to state which new items are included as opposed to being measured separately. To illustrate the last point: when describing cutting a new opening, state whether or not the new lintel and flooring in the opening are to be included in the price. Often it is advisable actually to measure and describe the items when on site so that one is aware of the existing conditions. Usually any locations given should relate to the existing building rather than the proposed layout, thus enabling estimators to locate them easily when pricing on site.

Surplus materials arising from their removal become the property of the contractor who is responsible for their disposal. If the employer wishes to retain any of the surplus materials or re–use them in the works then special mention has to be made of this in the bill. If it is felt that any of the surplus materials may be of use or value, then the contractor could be requested to price their removal with a separate amount for credit, with a proviso that the building owner reserves the right to retain the materials allowing the contractor the credited amount, if any, plus say 10 per cent. This is probably best catered for in the bill by including two cash columns against the items, one being headed credits.

Shoring and scaffolding incidental to demolition, repair and alteration work is deemed to be included together with making good any damage caused by the erection or removal. However, shoring of structures which are not being demolished is measured as an item. Any temporary roofs and screens required are included as items with a dimensioned description giving details of weather and dustproofing if required. Details must be given of any toxic or special waste likely to be encountered during the work.

PREAMBLES

Inclusion of special preambles in the bill for this type of work is necessary to advise the contractor of any restrictions or other matters such as phasing of the work. Furthermore, requirements common to several descriptions can be included as preambles in order to reduce the length of the descriptions. For example, 'cutting openings is to include for quoining up jambs with brickwork toothed and bonded in gauged mortar to the existing work.'

DEMOLITIONS

Demolitions are divided into three main categories as follows:

(a) demolishing all structures
(b) demolishing individual structures
(c) demolishing parts of structures (excluding cutting openings and removing finishings etc.)

All demolition work is measured as items and sufficient information must be given in the description to enable identification and the levels down to which the demolition is to be carried out. If any particular methods of demolition are required or any restrictions are to be imposed, for example, on the use of explosives, then these have to be mentioned. Particulars have to be given of any services to be diverted, maintained or sealed off temporarily. Making good parts of the structure remaining and any materials to be stored for re–use or retained by the employer have to be detailed. Breaking up parts of the building below the ground are usually left to be included as 'extra over' the excavation items in the measured work.

ALTERATIONS - SPOT ITEMS

This work includes removing fittings, fixtures, plumbing, engineering installations, finishings and coverings. Also included is cutting openings and recesses, cutting back projections, cutting to reduce thicknesses and filling openings. None of the work included in this section should be in preparation for bonding or renewal, which is covered by separate rules mentioned below. All work is measured as items and described giving details or a dimensioned description. Care must be taken to ensure that the item can be identified and details given of any making good of the structure or finish required and extending the finishings. If any new work is included then its description must be equal to that required in the SMM for the measurement of new similar work.

REPAIRING/RENOVATING CONCRETE/BRICK/BLOCK/STONE

Cutting out and replacing concrete, brickwork, blockwork, or stonework may be measured as a superficial, a linear or an enumerated item as appropriate; each with a dimensioned description. In the case of concrete, details have to be given of the concrete stating whether plain, reinforced or gun applied, together with any treatment to the reinforcement and bonding. Formwork and making good to match the existing are deemed to be included. In the case of brickwork, blockwork and stonework, details of the materials have to be given together with the method of bonding.

Repointing brickwork, blockwork or stonework is measured superficially with information in the description of such matters as size and depth of raking out, type and mix of pointing, bond and size of bricks etc. Rules are also included in the SMM for resin or cement impregnation and injection, removing stains, cleaning surfaces, inserting wall ties, re–dressing to form new profiles and artificial weathering.

CHEMICAL DAMP PROOF COURSES TO EXISTING WALLS

The provision of these is measured as a linear item describing the method and type of chemicals, the centres for drilling holes, details of finishes to be removed and whether to brick, block or stone. Making good to holes and finishes after injection is deemed to be included.

REPAIRING/RENOVATING METAL AND TIMBER

This work, which also includes refurbishment, is either measured as a superficial, a linear or an enumerated item as appropriate. The work is described by means of a dimensioned description and must give all the necessary information including identification of the work to be explored, prepared and executed, together with any associated works required.

FUNGUS/BEETLE ERADICATION

The treatment of existing timber is either measured superficial, linear or enumerated as appropriate and accompanied by a dimensioned description. A description of the treatment is given together with identification of the work to be explored, prepared and treated.

ADDITIONAL RULES RELATING TO WORK TO EXISTING BUILDINGS

As mentioned above these additional rules only apply to certain sections of the SMM and are used when preparing for bonding to or renewal of the item. With the exception of services the rules include such items as bonding and jointing new to existing; stripping off, removing or taking down; raking and curved cutting and cutting holes. As far as services are concerned the rules include such items as breaking into existing, jointing to new, stripping out, provision of temporary services and testing.

Taking–off List	**SMM Reference**
Demolitions and shoring:	
Demolishing all structures	C10.1
Demolishing individual structures	C10.2
Demolishing parts of structures	C10.3
Permanent shoring	C30.4
Temporary roofs	C10.6
Temporary screens	C10.7
Breaking out obstructions in excavations	D20.4
Breaking out pavings in excavations	D20.5
Alterations – spot items:	
Removing fittings and fixtures	C20.1
Removing plumbing and engineering work	C20.2
Removing finishings	C20.3
Removing coverings	C20.4
Cutting openings or recesses	C20.5
Cutting back projections	C20.6
Cutting to reduce thickness	C20.7
Filling in openings	C20.8
Temporary roofs	C20.9
Temporary screens	C20.10
Repairs and renovations:	
Cutting out defective concrete and replacing	C40.1
Cutting out defective brick, block or stone	C40.3
Repointing brick, block or stone	C40.4
Resin or cement impregnation	C40.2
Removing stains from concrete, brick, block or stone	C40.5
Cleaning surfaces of ditto	C40.6
Inserting wall ties in brick, block or stone	C40.7
Re-dressing stone etc., to new profile	C40.8
Artificial weathering to concrete, brick, block or stone	C40.9
Chemical damp proof courses:	
To existing brick, block or stone walls	C41.1

Taking-off List (cont.)	**SMM**	**Reference**
Cladding/covering, waterproofing, linings/ sheathing/dry partitioning, windows/doors/stairs and surface finishes:		
Bonding/jointing new to existing	Additional	rule 1
Stripping off/removing/taking down	Additional	rule 2
Making good disturbed work	Additional	rule 3
Jointing bonding new to existing	Additional	rule 4
Raking and curved cutting	Additional	rule 5
Cutting holes for ducts and pipes	Additional	rule 6
Drainage work and mechanical services:		
Breaking into existing pipes and ducts	Additional	rules 1/2
Jointing new pipes and ducts to existing	Additional	rules 3/4
Stripping out installations	Additional	rules 5/6
Provision of temporary services	Additional	rule 7
Stripping off insulation	Additional	rules 8/9
Testing, etc.	Additional	rule 9
Electrical services:		
Work to existing electrical installations	Additional	rules 1–11

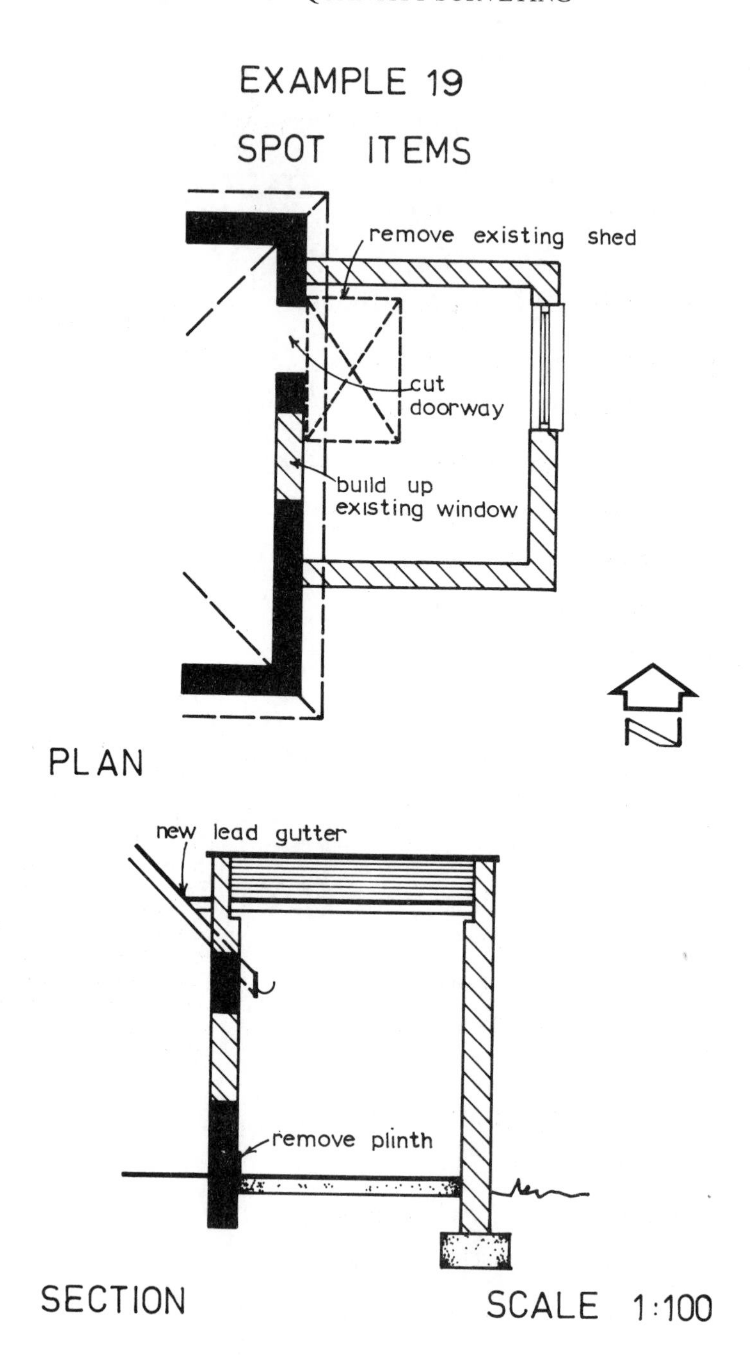
EXAMPLE 19
SPOT ITEMS
remove existing shed
cut doorway
build up existing window
PLAN
new lead gutter
remove plinth
SECTION
SCALE 1:100

FIGURE 38

SPOT ITEMS

Spot Items 1.

			Description	Notes
			Pull down to ground level the shed adjacent to the East wall of Kitchen about 1900 x 1300 and 2100 high, including making good remaining brickwork and facings	Description indicates location approximate size and work required in making good. C10.2.1.1.3
1.90 1.30			Break up and clear away Concrete floor and hardcore bed under (Assumed 150 Concrete, 150 hardcore)	Looking at the floor neither the surveyor or the estimator can guess the thicknesses. An assumption is made and if its subsequently found that the assumed thicknesses were significantly different, the item may have to be adjusted in the final account. If of course an original drawing was available, as sometimes is the case, then no assumptions are needed and the actual thicknesses can be stated.

ALTERNATIVELY

			Description	Notes
			Break up and clear away Concrete floor and hardcore bed under last (About 2.5 m²) (Assumed 150 Concrete, 150 hardcore)	C 10.3

Spot Items 2.

Take out and clear away window 1200 x 1200 in East wall of Kitchen together with frame and stone sill under. Fill opening with common bricks in cement mortar 1½ brick thick, including cutting and bonding to jambs and pinning up to soffit. Plaster both faces and adapt and extend wall tiling on Kitchen side. Make good all work to match existing.

A comprehensive description of the actual job required to be done allows the estimator looking at the building and reading this description to get the overall picture in his mind.

A comprehensive description such as this complies with the requirements of SMM. However if there are many windows to be filled all the same size, consideration should be given to measuring the filling and the plastering as separate items making clear in each case that they are adjustments to the spot item. It could simplify the pricing and help the Contractor by isolating the sub-trade work.

C 20.8

Cut opening 850 x 2050 in 1½ brick wall adjoining last window for new door opening, cut out for and insert concrete lintol (measured separately), face up jambs, make out flooring and skirting to new opening and make good plaster up to new linings.

C 20.5

Spot Items 3.

			Door. 750 Frame. 2/50 = 100 Bearing 2/150 = 300 1150	
1			327 × 140 Precast Concrete (20N/mm²) lintol 1150 long reinforced with and including 3 Nr 12 mild steel bars	Alternatively the lintel could have been measured with the new door. F 31.
			[New door to take.	Reminder to ensure new work in connection does not get forgotten.
			Cut back the projecting brick plinth 25 × 300 on the East elevation for a length of about 4400, including making good both ends up to new wall	C 20.6
			Take down eaves gutter and fascia for a length of about 4400, including making good both ends of fascia and providing stop ends to gutter	In measuring this item on site the surveyor would have checked that no work was required to the down pipes. i.e. the rainwater can still get away. C 20.2

Spot Items 4.

Strip roof coverings and cut back projecting rafters for a length of about 4400, prepare 1½ brick wall for raising, including making good roof up to new gutter

C 20.4

Eaves filler & gutter to take.

Reminder as before.

19 Taking–off a complete building

This chapter makes various suggestions, chiefly regarding procedure, for the guidance of the taker–off in dealing with a complete building. It must not be forgotten that the methods of different individuals and the customs of different offices vary in the organisation and carrying out of this work. The fact that a particular suggestion is made here does not mean that it is universally adopted, nor does it preclude the use of an alternative.

RECEIPT OF THE DRAWINGS

As the drawings are received from the architect they should be stamped with the date and entered on a list showing the drawing number, title, date of preparation, date received and the number of copies. This list forms a useful reference and enables one to see when any revisions to previous drawings were received. A check should be made with the architect to ascertain whether any further drawings are under preparation which are likely to be received before the bill of quantities is completed. At the completion of the taking–off each drawing used should be stamped 'used for quantities' which avoids confusion later on as to whether or not subsequent revisions were incorporated in the bill.

PRELIMINARY STUDY OF DRAWINGS

Before any dimensions are written at all the senior taker–off should look over the drawings and study the general character of the building.

A check should be made to ensure that all floor plans and elevations are shown and that the position of the sections is marked on all the plans. A more detailed inspection should be made to ensure than windows, doors, rainwater pipes etc., shown on the plans are also shown on the elevations and vice versa. If any of the drawings have been prepared by specialist engineers they should be compared with those prepared by the architect to ensure that there are no discrepancies in the layout or dimensions. If overall dimensions are shown these should be checked against the total of room and wall measurements. If overall dimensions are missing these should be calculated and marked on the plans, also all projections on external walls should be dimensioned. If this is done the calculation of the mean girth and of the perimeter of the external walls is simplified. It is an advantage to dimension every room on the plans except where a series of rooms are obviously all of the same dimension in one direction. A little time spent in this preliminary figuring of the drawings obviates the possibility of inconsistency in dimensions. Where more than one taker–off is employed the result of this work must be communicated to each, and their drawings marked accordingly. Figured dimensions must always be followed in preference to scaled and any dimensions which can be calculated from those figured should be worked out. A larger scale will usually override a smaller scale drawing, except when the smaller is figured and the larger is not.

QUERIES WITH THE ARCHITECT

Preliminary inspection of the drawings may also give rise to a number of queries to be raised with the architect either relating to missing information or discrepancies. Questions settled at an early stage saves interruption to the taking–off and consequently increases productivity. Queries should be listed on the left–hand side of a sheet of paper with the answers, when received, shown opposite on the right-hand side, together with the date and source of the answer. The quantity surveyor, however, must accept some responsibility in making decisions and many architects are quite ready to confirm such decisions. These query lists should from time to time be sent to the architect either for completion or confirmation. It is important not to keep worrying the architect with queries particularly over the telephone or the Fax.

Early preparation of schedules of finishings, windows and doors often exposes missing information and they also provide a useful reference for the whole taking–off team. If materials are shown for which full up to date information is not available in the office it is advisable to ask manufacturers for literature and advice at an early stage. If the survey of the existing site does not show levels or those that are shown are insufficient, it is prudent to take, or have taken, a grid of levels over the site. These levels will prove invaluable when measuring earthworks and are essential for reference later if there is any question of re–measurement.

INITIAL SITE VISIT

Before actually starting on the measuring it is advisable to make an initial visit to the site. If nothing else such a visit allows one to get the feel of the job and be able to visualise later the various site references when they are encountered on the drawings. A further visit or visits will be necessary when the measuring has progressed to pick up the spot items (see Chapter 18) and site clearance items (see Chapter 17). The initial visit will give an indication of what will be required in due course; in this respect other items to be noted include such matters as boundaries generally (state and ownership of fences, walls, gates etc.) the existence or otherwise of overhead and underground services, means of access, adjoining buildings (historic, civic, defence etc.) and, all matters that will be required for the subsequent drafting of the preliminary bill (see Chapter 22). In addition, this visit is an opportunity to take the levels referred to above and also to examine and note any trial holes that may have been dug and are open for inspection. A photographic record of the site or specific parts of it may prove useful as a reminder of site conditions as existing.

WHERE TO START

If all drawings are completed and available the taker–off will probably follow the order of sections given in Chapter 5 or such other order as may be the custom of the office. This order will be seen to follow more or less logically the order of erection of the building, except that certain special work and services are dealt with at the end. It may however be that the design of, say, reinforced concrete foundations is not completed, and therefore measurement cannot begin as usual with foundations. In such a case a point such as damp proof course level must be selected from which to measure the structure, the work below that level being measured at a later stage when the necessary information is available. It may be that 1:100 scale drawings have been received, but 1:20 details are to follow. In such a case internal finishings could probably be taken first, as the measurement of these is dependent generally upon the figured dimensions given on the 1:100 scale, which is to be presumed will not be altered. Moreover, these sections give the taker–off a good idea of the plan and general nature of the building. Specific forms of construction e.g. a steel or concrete frame will require a special sequence of measurement to be devised.

ORGANISING THE WORK

Where several takers–off are involved in the measurement of work and a deadline has to be met for the completion of the bill it is important that the team leader organises the work carefully. To enable progress to be monitored a schedule should be prepared showing the sections of work to be measured with the name or initials of the taker–off responsible, together with the target and actual dates of commencement and completion of the work. Clear instructions must be given to each member of the group, carefully defining the extent of the work to be taken in each section. Proper arrangements should be made for the collection of queries for the architect and these should be edited by the team leader. Proper supervision should be made for junior staff involved and due allowance made for staff leave and their commitment to other work.

ACCURACY

There may be a temptation, especially to those not sure of their ground, to take measurements on the full side, and so cover themselves against possible claims for deficiency. Such measurements are very definitely to be deprecated, as it will be realised that, if measurements were on an average 2½ per cent, or perhaps 5 per cent, full, a very substantial amount would be added to the tender. Where figured dimensions are the basis the measurements taken should be exact, but where they are scaled there is some excuse for measurements slightly full, as there may be some shrinkage of the paper and the lines on the drawing are not usually drawn sufficiently precisely to scale nearer than, say, 50 mm on a 1:100 scale plan with any certainty. It will be seen therefore that the care with which the architect's drawings are prepared may affect the quantities, and that measurement from fully figured drawings must necessarily be more accurate. Furthermore, the quantity surveyor owes a duty to the client to prepare an accurate bill of quantities and it could be regarded as negligent not only to under measure but also to over measure.

When working from figured drawings it may be found that three places of decimals are in some cases indicated. In setting down the dimensions the figures should normally be taken to the nearest two places of decimals, the excess or loss in measurement thus obtained being usually very slight in proportion to the whole, particularly when the total bill quantity is usually given to the nearest metre. Three places may, however, be desirable when setting down waste calculations.

In short, the surveyor must aim at giving in the bill a representation as accurate as possible under the circumstances of the building work in question. There is, however, a sensible limit to the degree of accuracy to which one should measure and more experienced takers–off often take the value of the item into account when deciding on the degree of accuracy required.

NUMBERING THE DIMENSION SHEETS

All dimension sheets should be numbered as soon as possible. For those in traditional form some surveyors number each column, others each page, this being a matter of individual preference. If the numbering is done at the earliest opportunity, it will minimise the risk of a page being mislaid and overlooked. If sheets are inserted after numbering, they can be numbered with the last number and a suffix a, b, c, etc., but it is essential that if, say, sheets numbered 31a, 31b, 31c and 31d are inserted, a note should be put against the number 31 '31a – d follow' otherwise the inserted sheets might be lost and not missed. A similar note should be made if gaps occur in numbering.

In any case it will be found of value if the taker–off numbers the pages at the top, according to the section they represent, e.g. floors 1, roofs 6, plumbing 9, etc., irrespective of any other system of numbering. These pages can be kept in order awaiting the main sheet numbering which may follow on from another section. It is further of value in that on opening the dimensions at any page one has an immediate indication of which section is measured. Special rules apply to the method of numbering of 'cut and shuffle' sheets as explained in Chapter 3.

CROSS–REFERENCES

The taker–off should try to ensure that dimensions are clear to others, as it is quite possible that, when variations on the contract have to be adjusted, someone else will be entrusted with the work, and will have to find their way about the dimensions. If the taker–off wishes to insert an item and finds that there is no room for it in its proper place and is forced to put it elsewhere, a reference should be made on the dimensions in the correct place so that the item can be found and vice versa. Any further cross–referencing should also be made which may be of use later. It has been known for jobs to be postponed for a year or two after tenders were received, and even the taker–off will then need some references and notes to refresh the memory.

CLEARNESS OF THE DIMENSIONS

Besides the use of cross–references, a good deal can be done to make the dimensions clear by the manner in which they are set down. It has already been pointed out that a regular order of length, width and depth (or height) should be maintained in writing down the dimensions, and even when it may be difficult to determine which is length or width a consistent order should be kept. In measuring areas of floor finishings, for instance, the dimensions horizontal on the plan could be put first, followed by the vertical ones. Calculations should be made wherever possible on waste and not mentally and timesing should be made consistently. For instance, in measuring six doors each with four squares of glass all timesed for two floors the dimensions should be timesed thus:

Timesing	Dimensions	Squaring	Description
2/6/	1		50mm Wrot softwood four panel door etc.
2/6/4/	0.23 0.45		6mm Clear sheet glass to wood etc.

the last item not being written as follows:

Timesing	Dimensions	Squaring	Description
2/4/6/	0.23 0.45		6mm Clear sheet glass to wood etc.

It will be realised that when timesing becomes complicated it will help considerably in tracing items if the method of timesing is consistent. In the example above, the outer timesing represents the floors and the next the number of doors, and if the order is reversed in the middle of a series there could be confusion. The use of coloured pens in timesing or dotting on to represent different floors or sections of the work will be found to help considerably in tracing dimensions later.

SEQUENCE OF MEASUREMENT

It is advisable when measuring to follow the same sequence in different parts of the work. For example, in collecting up the girth of the external walls of a building it is a good plan, as suggested above, to work clockwise, starting say, from the top left–hand corner of the plan. If this is done consistently it will assist in reference later, when perhaps the length of a particular section of wall has to be extracted from a long collection. If a particular sequence of rooms has been adopted in measuring ceilings, the same sequence should be used for wall finish, skirtings, floors, etc. In this way, if all the finishings of a particular room are to be traced, it will be known in which part of each group the relative dimensions are to be found.

HEADINGS

The use of headings in the dimensions will further help in future reference. Apart from the taking–off section heading already suggested for each page, sub-headings or 'signposts' should be clearly written wherever possible, and these will stand out if underlined. The sequence of measurement in the section will then easily be followed by a glance at the sub–headings. Special sub–headings will also be necessary where it is required by the SMM that a group of dimensions are to be billed together, as, for instance, in the detailed measurement of manholes. These bill headings should be written across the dimension and description columns and underlined to distinguish them from signposts and the end of the work clearly defined by a note such as 'end of manholes'.

NOTES

The making of notes by the taker–off on the dimensions is of the utmost value. Such notes are usually of one of three kinds:

(1) Explanatory for use in writing the specification or adjusting variations.
(2) For reference before the taking–off is finished, e.g. notes of items to measure, or queries to be settled.
(3) Instructions for preparing the bill

If a specification is not available prior to taking–off then the explanatory type of note should include memoranda for the specification, such as the rules followed in measurement of lintel reinforcement, or, say, in the case of a plate bolted down to brickwork, the spacing of bolts assumed. Such notes will save the specification writer from spending time trying to see on what principle reinforcement is measured or how far apart bolts were taken, and even if the taker–off writes the specification such notes will be found of value, particularly writing a month or two after the taking–off is finished. They will also be found of value in adjusting variations, especially if this is done by someone else. This type of note should be written at the side of the description column on 'waste', and is best separated by a line or bracket to prevent confusion with descriptions.

The second type of note, made for reference before the taking–off is finished, is necessary when perhaps some point must be referred to the architect, or for other reasons something cannot be finally measured. As mentioned previously a list of such queries should be compiled. 'To take' notes are entered in dimensions when a taker–off dealing with one section feels that an item, although arguably within the work being measured, is better taken with another section e.g. tile splashbacks to sinks taken with the plumbing rather than with the finishings or vice versa. These 'to take' notes should be written clearly in the dimensions and collected together before the bill is written and a check made to ensure that all items listed have been measured. Memoranda, too, should be made at the end of the day's work of anything unfinished which, the train of thought being broken, might be forgotten. It is much safer to make notes of such matters on the dimensions than to trust to memory, and these notes enable the work to be carried on in the case of unexpected absence. Such notes as these are sometimes written in pencil to be erased when dealt with, or they may be written in ink across the description column in such a way that they cannot be missed. It is a well–known fact that the subconscious mind continues to work long after the conscious mind has turned to something else, and the subconscious thoughts have a way of pushing themselves forward suddenly at most unexpected times. If anything of importance should occur to the taker-off in this way during leisure hours – something forgotten or something which could be improved – a point should be made of writing it down, or the thought may vanish as quickly as it came.

The third type of note is that which is an instruction to the writer of the bill used generally where a correction or alteration can be more easily made by a general instruction than by writing down new dimensions or descriptions. These notes, too, must be written in such a way that they will not be missed. An example of this type of note would be such an instruction as 'alter all 25 mm shelving to 32 mm'.

DRAWINGS

As already mentioned, figured dimensions on drawings should be used in preference to scaling. Naturally, where there is a large discrepancy between the two, one would compare with other dimensions given, or, if this fails to produce a solution, with the architect. Care must be taken to use the correct scale when taking measurements from drawings, particularly where a variety of scales has been used. As items are measured from drawings it is often beneficial to loop through the applicable written notes and perhaps colour in the work on the drawing. An examination of the drawings thus marked, at the end of measurement, will soon reveal any items not measured. Not all measurable items are shown on drawings, particularly if the drawings are incomplete or in any case labours such as surface finishes to concrete, but the method should avoid any major items being forgotten.

THE SPECIFICATION

If a specification is supplied, it should be read through cursorily first, not with the idea of mastering it in detail, but rather with a view to getting a general idea of its contents and knowing one's way about it. One would then normally study for example the excavation, concrete work and brickwork more closely before beginning the taking-off for the foundations. It is useful when the taking-off is well forward to go through the specification and run through in pencil all parts that have been dealt with, not of course paragraphs which will form preambles to the bill or such descriptions which have not been repeated in the dimensions. If this is done it is unlikely that anything specified will be missed. The specification is not, under the usual JCT form of contract, a contract document, but if the bill and specification both form part of the contract they will have to be integrated together. In such cases the wording of the bill descriptions can be reduced by reference to the specification particularly in a co-ordinated document. When the specification is a contract document, the preambles and the preliminaries in the bill can be similarly reduced in length.

More often than not, however, no formal specification is available. Brief specification notes may be supplied, and further notes must be made of verbal instructions which will be given by the architect and which will form the nucleus of the specification, to be added to from time to time as queries are raised. These notes must be supplemented by the surveyor's own knowledge of the architect's usual practice or by ideas of what is reasonably required. In making decisions on matters of specification the surveyor should bear in mind that what is theoretically correct is not necessarily the most practical or economic solution. In the case of timber, for instance, not only the difference between basic and finished sizes referred to in Chapter 9 should be borne in mind but where possible sizes which are readily available in the market should be specified. The proliferation too, of minor variations of the same article does not lend to economy, especially in terms of joinery where machines have to be reset for each different moulding, in fact the SMM in the case of door frames calls for the incidence of repetition of identical items to be referred to.

PC ITEMS AND PROVISIONAL SUMS

For various reasons it is not always possible to define finally, when quantities are being prepared, everything necessary for the completion of the building. For instance it may be necessary for the architect to select certain articles such as sanitary appliances, ironmongery, etc., in consultation with the client, and the details of these may very well not have been considered at the early stage when tenders are being obtained. It is not unusual, therefore, to put in the bill 'prime cost' ('PC') sums for these items which the estimator will include in the tender for goods to be obtained from a supplier, but which are subject to adjustment against the actual cost of the articles selected. The contractor is given the opportunity in the tender to add for profit to each of these items.

Further, it may be necessary to employ certain firms to carry out work which a builder does not do in the normal course of business, or which it is considered is better done by a specialist e.g. asphalt, electrical work, heating and hot water services, lifts, etc. If possible, estimates for these works will be obtained from specialist firms and PC sums included in the bill of quantities, which the estimator will include in the tender. The firms eventually employed for these works will be sub-contractors of the general contractor, commonly known as domestic sub-contractors. The main contractor retains general control and responsibility. Again the main contractor is given the opportunity in the tender to add for profit and also for attendance as described below.

With regard to PC sums, as the JCT forms of contract allow different cash discounts to the main contractor for sub-contractors and suppliers, it is important to keep them quite distinct.

Provisional sums are sums included for general contingencies and work to be done by the general contractor which is at the moment uncertain. How PC and provisional sums are incorporated in a bill of quantities is fully dealt with in Chapter 22.

ATTENDANCE

An item of general attendance on nominated sub-contractors is required by SMM A42.1.16 to be given in the preliminary bill. This is a general item covering all sub-contractors and the meaning of general attendance is defined in coverage rule C3 of SMM A42. It covers such items as use of erected scaffolding, temporary water and lighting supplies, messing facilities etc., and items of a general nature applicable to all sub-contractors.

Other or special attendance referred to in SMM A51.1.3 will differ according to the nature of the work and must be considered for each item. Examples contained in this rule include hardstanding, power, storage etc. The conditions accompanying the quotation (often small print on the back) can be helpful in drafting this item.

APPROXIMATE QUANTITIES

Work to be carried out by the general contractor, which may be uncertain in extent, can also be provided for by means of approximate quantities, i.e. by measuring work in the normal way, but keeping it separate in the bill and marking it 'approximate'. For instance, the foundations of a building, where the nature of the soil is uncertain, may be measured all as shown on the drawings and additional excavation, brickwork, etc., measured separately and marked 'approximate' to cover any extra depth to which it may be necessary to take the foundations. It is thus made clear that adjustment of these quantities on completion of the work is anticipated. Alternatively if there is a considerable amount of such work it may be contained in a bill of approximate quantities. If the work cannot be described adequately it is to be included as a provisional sum (SMM GR10.2).

DIAGRAMS

In certain circumstances the SMM calls for dimensioned diagrams to be included as an adjunct to the written description. These diagrams are defined in SMM GR 5.3. They will be either extracts from the drawings or drawings specially prepared for incorporation with the written description or on the facing page. Alternatively the diagrams are collected together and printed on one or more sheets at the end of the bill, each diagram being given a number and being referred to by that number in the body of the bill. Diagrams should be marked 'plan', 'section', etc., and be drawn to scale, the scale being indicated or the dimensions figured.

Apart from the above, sketches will be made in the dimensions or on separate sheets to work out points of construction not detailed or to supplement the architect's details. If possible, these should be made on the dimension sheets where the particular work is measured, on a spare space on the drawings, or if made on separate sheets they should be carefully preserved to show at a later stage what the taker–off had assumed when measuring. Any details of importance should be confirmed with the architect.

MATERIALS

The taker–off must have a thorough knowledge of the materials being used, and should, where possible, if a material is unknown to him make a point of seeing samples and studying the manufacturers' catalogue or leaflet, so that the limits and (perhaps optimistically stated) capabilities of the material are known. The handling of the material and study of any literature describing the materials and the way it is to be fixed in position will often assist in the measurement of the work or in the framing of a proper description.

INSERTION OF ITEMS

It quite often happens that the taker-off must go back on the work and make alterations or insert additional items, either because items have been forgotten or because revised details have been received, as described above. Such alterations and additions should wherever possible be made in the proper place in the dimensions, so that when the dimensions are referred to at a later stage everything can be found where expected. Once more the importance of plenty of space must be emphasised, and on work of any size it will be found of advantage to start each section of the taking-off on a new sheet, leaving any odd blank columns for later use if necessary. Further, if the dimensions are kept in subsections or groups with a definite gap between, these gaps will also be found of use to insert in their proper place any dimensions which may be an afterthought. If it is found impossible to insert an item or group of items in its proper place, a place must be found for it elsewhere and proper cross-references made in both places, as described previously.

DECISION ON DOUBTFUL POINTS

A thorough knowledge of the SMM, the Measurement Code and a study of all the published textbooks will still leave occasions when the taker-off must make decisions for himself on the method of measurement or extent of descriptions for items not covered in the SMM. When a rule of measurement is originated in such a way it is often advisable to insert the method used as a preamble clause in the bill. When making such decisions the taker-off should have one main consideration: what will best enable the estimator to understand (not merely guess) the work involved and enable it to be priced quickly and accurately.

20 Abstracting

OBJECT AND NATURE OF THE ABSTRACT

In splitting up the building into its constituent parts for measurement the taker–off follows a systematic method but there has to be a repetition of the same description in different parts of the dimensions. Moreover, in the same group of dimensions there will be items that will appear in different sections of the bill, which usually is divided according to the sections of the SMM. The function of the abstract is to collect similar items together and to classify them primarily into the SMM sections and subsequently according to certain accepted rules of order and arrangement, thus being in a suitable sequence for writing the bill.

The abstract is prepared by copying the descriptions and squared dimensions from the taking–off onto large sheets of paper in a tabulated form as nearly as practicable in the order of the bill as described in Chapter 21.

As stated previously the need for an abstract is eliminated by the use of the 'cut and shuffle' system. In that system the principles behind the abstract are retained, but the order is effected by having each item on a separate slip of paper and sorting them into the correct order.

Whilst most offices have adopted the 'cut and shuffle' or a computer system of sortation, there are occasions when it may be more expedient to prepare an abstract. This chapter gives an outline of the traditional form of abstract.

NUMBERING THE DIMENSION PAGES

When the taking–off is complete the dimension pages should be examined to ensure that they are numbered consecutively and that no sheets are missing. If sheets have been numbered temporarily they should be given permanent numbering which will be used for referencing items on the abstract.

SQUARING THE DIMENSIONS

A check is made of the waste calculations and their correct transfer to the dimensions. The dimensions are then calculated or 'squared' and where bracketed together are totalled, subtracting any deductions which follow immediately. The 'squarings' and 'casts' are checked and ticked, preferably in red.

ABSTRACT PAPER

The abstract paper is usually a sheet of A3 ruled in columns about 25 mm wide with a blank strip across the top for headings and folded vertically down the centre. The sheets should be opened flat and used like an unbound book. The front of every sheet is stamped with the name of the contract and headed, usually in the top left–hand corner, with the names of the bill section and sub–section. Both sides of the paper are used and entries are made on the sheet using one pair of columns for each item. The order in which items are entered on the sheet is from the top left–hand corner on the front of the sheet working down each pair of columns to the bottom right–hand corner, using the back of the sheet in the same way.

SETTING OUT THE ABSTRACT

Before setting out the abstract an attempt should be made to visualise the form of the bill. To do this the dimensions should be examined and the taker–off consulted to ascertain the general nature of the work. It should also help the abstractor to look at the drawings to visualise the work and form an idea of the general character of the building. The actual spacing out of items on the abstract is largely a matter of experience, although a bill for a previous similar job may form a useful guide. Experience in abstracting may be gained by going through the dimensions and picking out items for one section of the bill. This is obviously slower than abstracting each item as it appears on the dimensions which has the added advantage of enabling the abstractor to follow the taking–off and point out possible omissions or errors to the taker–off. Traditionally abstracting formed a useful training for potential takers–off, although some staff used to create a successful career in abstracting and billing, i.e. 'working–up'.

ENTERING ITEMS IN THE ABSTRACT

Examples follow of how items from the dimensions may be entered into the abstract.

10/	3.95	39.50	50 × 200 Sawn sw in floors
2/	3.95	7.90	75 × 200 Ditto
	1.24	1.24	
			9.14
2/	3.95	7.90	Deduct 50 × 200 Ditto
2/	0.75	1.50	
			9.40

L/Sawn sw in floors	
50 × 200	75 × 200
39.50 (96	9.14 (96
Deduct	
9.40 (96	

L/50 × 200 Sawn SW in	floors
39.50(96	Deduct
	9.40(96
L/75 × 200 Sawn SW in	floors
9.14(96	

Usually one pair of columns is used on the abstract paper for each item and, as shown in the second example, the right hand column of the pair is used for deductions. The unit of measurement of each item should be shown by the letter C, S, or L and the description is underlined before the quantities are entered. Each quantity is referenced with the page number separated by a bracket. If further sizes of timber occur later in the dimensions the size headings, as shown in the first example, could be extended further across the columns to the right (or left). Plenty of space should always be allowed between items on the abstract and a maximum of three descriptions vertically per pair of columns should be the aim.

RUNNING THROUGH THE DIMENSIONS

As each item is entered on the abstract, it is 'run through' on the dimension sheet with a vertical line, the top and bottom of which should be clearly defined with a bar. This enables the abstractor to pick out the items from the dimensions in any order and to leave no doubt as to which items have been abstracted. An example of lining or 'running through' follows:

	20.00 3.00		2ct plast to bk wls & 2ce emuls pt to plast wls
2/	3.00	6.00	Extl ang to wl plast

Another way of 'running through' used by some surveyors and of special value when descriptions are cramped is as follows:

	30.00 3.00	90.00	2ct plast to bk wls & 2ce emuls pt to plast wls & Deduct 2ct plast to block wls & Deduct 2 cts gloss pt to plast wls

The items entered are 'run through' with a diagonal line, the vertical line being drawn through all four items when the last item in the group is abstracted. This last item to be abstracted need not be the last item as it appears on the dimensions since a small amount of looking forward on the dimensions may save some switching from one section to another in the abstract.

ABBREVIATIONS

It has already been pointed out that descriptions on the dimension sheets when using an abstracting system of working–up can be considerably abbreviated; these same abbreviations being used in writing the abstract. The items should be copied as written by the taker–off, the abstract really being a rearranged form of the worked–out dimensions and descriptions. If, however, any description is particularly long it can be shortened on the abstract with a note referring back to the appropriate dimension page number. This method should be reserved for very long descriptions only as the necessity for constant reference back to the dimensions when writing the bill may waste more time than is saved in abstracting.

ADVANTAGE OF SPACE

The abstractor should always aim to take too much space rather than too little. It is even better to waste a little time in copying out portions of the abstract which have become too crowded rather than to risk the chance of confusion from items getting mixed up. The abstractor should be looking out for the possibility of overcrowding and decide as early as possible if any part is to be rewritten. If the rewriting is done after the abstract has been checked partly, great care must be taken to repeat the checking.

ESSENTIALS OF A GOOD ABSTRACT

Apart from the need for space emphasised above, a good abstract must have numbers neatly written with digits carefully lined up under one another so that no confusion will arise in totalling the figures. References to the dimensions must be accurate as these enable bill items to be traced back readily to the dimensions.

ITEMS BILLED DIRECT

Where items occur in a group in the dimensions which will be grouped together similarly in the bill, there is often no necessity to copy them out onto the abstract. For example, all the work in drainage will probably be measured together and will appear in the bill in one section. A note can be made at the appropriate point in the abstract to bill direct (or B/D) pages 365 to 369 inclusive. When the biller reaches this point on the abstract the bill will be written direct from the dimension sheets picking out the items in bill order and if necessary creating a mini–abstract as mentioned below.

ABSTRACTING ON THE DIMENSION SHEETS

It is sometimes possible to prepare the equivalent of an abstract on the dimension sheets, when perhaps, say, some simple plumbing work can be billed direct, or when measurements are taken 'trade by trade'. Such collections of similar items will be done in the same way as on the main abstract, and page references against transferred dimensions are equally important in order to save time when tracing them in the future. These mini–abstracts can either be done on any suitable space available on the dimension sheets or the totals alone from subsequent items can be transferred to and added to or deducted from the quantity of the first similar item. As items are transferred they should be 'looped through' to show that they have been dealt with, the normal lining through being used on the 'master' items as they are billed.

CHECKING THE ABSTRACT

Before the checking of the abstract is commenced the pages should be numbered in the top right–hand corner to prevent any chance of a sheet being mislaid. If the abstract is incomplete the pages should be numbered temporarily in pencil. In the process of checking, each item on the abstract is ticked, preferably in red, the tick being placed between the quantity and the page reference. As each item is checked a red line is run through the dimensions parallel to and similar to the line made by the abstractor. If diagonal lines have been used by the abstractor the checker's lines should be made in the opposite direction. The checking process is in fact a repetition of the abstracting, but instead of the item being written it is checked and amended if necessary. As always in any quantity surveying work figures to be corrected should be crossed through neatly and rewritten; alterations made by superimposing the correct figures are likely to be misleading and correcting fluid should never under any circumstances be used. Descriptions, particularly those containing figures, as well as dimensions and references should be checked carefully; again alterations being made by crossing through and rewriting. Any corrections made in checking should be cross-checked. On completion of the checking, the abstract should be inspected to ensure that all entries have been ticked. Similarly the dimensions should be examined to ensure that all items have been crossed through with checking lines. If during checking an entry cannot be found in the abstract it should be entered in red ink, the subsequent inspection of the abstract may reveal a corresponding unchecked item in another place. In this event one of the duplicated items must be deleted.

CASTING UP

On completion of the checking the figures in the abstract are totalled or 'cast up'. After casting, the totals of deductions are transferred to the bottom of the addition column and the subtraction made, the original deduction figure being looped through to indicate that it has been transferred. An example follows:

S/ 2 Ct plast to bk wls	
	Deduct.
1 432.00 (16.✓	
723.00 (26.✓	94.00 (64.✓
212.60 (27.✓	12.00 (67.✓
103.00 (29.✓	10.60 (72.✓
2 470.60 ✓	116.60 ✓
116.60 ✓ D.	
2 354.00 ✓	

All casts, transfers and deductions are checked and ticked as before.

REDUCING

Generally all items are billed in metres to the nearest metre and the resultant total from the casting is ready for transfer to the bill either ignoring decimal figures if below 0.5 or taking to the next whole figure if 0.5 or over. Some items, however, have to be billed in different units from the measurement e.g. bar reinforcement billed in tonnes to the nearest two decimal places. In this instance the total measured length is multiplied by the weight of the bar per metre and the result entered below the item on the abstract ready for transfer to the bill, as shown in the following example:

L -- dia steel bar, straight	
12 mm	10 mm
95.00 (46. ✓ 210.00 (51. ✓ 16.00 (54. ✓	152.00 (50. ✓ 200.00 (61. ✓
321.00 ✓	352.00 ✓
× 0.888 kg ✓	× 0.616 kg. ✓
= 285 kg ✓	= 217 kg ✓
= 0.29 t ✓	= 0.22 t ✓

The reductions are checked before the billing is commenced.

21 Bill preparation – general

PURPOSE OF THE BILL

Before considering the techniques of preparing the bill one should bear in mind the main purposes of the document. First and foremost the bill is used to assist contractors with the preparation of an estimate for tendering. The priced bill, which becomes a contract document, provides a basis for the valuation of varied work and of work completed for stage payments during the contract. In addition the bill should be of use to the contractor in the organisation of the work and to the surveyor to provide historic cost information.

NEED FOR DRAFT BILL

There are two main circumstances when it may be necessary to write a draft bill; firstly when an abstract of the dimensions has been prepared in the traditional way and secondly when a 'cut and shuffle' system has been used in which the descriptions are not written in full by the taker–off and therefore the bill cannot be typed directly from the sorted slips. Even when a 'cut and shuffle' system is used with full descriptions there is a considerable amount of editing to be done to the sorted slips in preparation for typing. Whichever system is used most of the remarks contained in this chapter apply.

PROCEDURE

The writing of the bill is theoretically copying out the descriptions and quantities from the abstract or sorted slips in the form of a schedule or list on paper ruled with cash columns for pricing, but in practice it is a good deal more than this. The taker–off may describe an item briefly if it is in common use and leave it to the billing stage for the full description to be compiled. In fact the term 'working–up' originated from the task of working–up or expanding the brief taking–off descriptions to proper bill items.

Furthermore, where several measurers are working on the same project, their descriptions must be co–ordinated. They cannot constantly be asking each other for the exact wording of the description for a particular item, and so the different takers–off will sometimes describe the same work slightly differently. The biller must therefore understand what is being written and see where different descriptions approach each other so closely that in effect they mean the same thing. Conversely the biller must be able to detect subtle differences in descriptions which may affect the price of an item.

DIVISION INTO SECTIONS

The bill of quantities has been traditionally divided up into work sections similar to those contained in the SMM. With the introduction of SMM7 the likelihood in the future is that the bill will reflect the main common arrangement sections of the rules. In certain countries it is customary to invite separate tenders for each trade, but in England, unless management contracting has been adopted, it is usual to invite tenders from a general contractor only. Where separate tenders are invited the need for separate bills for each trade is obvious, but even where tenders are obtained from a general contractor the division into work packages is of assistance in pricing and simplifies the estimator's task if any particular part of the work is to be sublet. Moreover, it is this first step in the subdivision of the bill which enables items to be easily traced. The sections into which the SMM is divided generally align with sub–contractor's work and any departures such as precast concrete bollards being contained in furniture/ equipment are soon recognised. Work to existing buildings, work outside the curtilage of the site and work to be subsequently removed have to be billed in separate sections (SMM GR7). Traditionally substructure or work up to and including damp proof course and external works are also kept in separate sections. These indications of location should assist the estimator in pricing and furthermore facilitate the valuation of, and measurement of possible amendments to, this work at a later stage. When a project consists of several buildings it is often desirable to provide a separate section in the bill for each one. Alternatively, as suggested in the MC, a multi-column analysis on the page facing the descriptions and quantities can be used if the construction of the blocks is similar. Besides the items actually transferred from the dimensions, the normal bill has other parts such as preliminaries and specification preambles to each work section describing materials and workmanship. These and other additional items are dealt with more fully in Chapter 22.

ELEMENTAL BILLS

For certain reasons it may be desirable to produce an elemental bill, in which the main divisions are design elements or constituent parts of the building, e.g. foundations, floor construction, windows etc. irrespective of the work sections of the SMM. The main purpose of such a bill is to assist a standardised system of cost analysis which may be adopted, particularly where buildings of a similar nature are to be repeated. The taking–off is done by elements and each element forms a separate section in the bill, the normal order of work sections being maintained within each element. Whilst in theory this type of bill should make estimating more accurate because the items are related to a particular part of the building, contractors who sublet work may have difficulty in collecting appropriate items together. Furthermore estimators may find similar items occurring in different elements and have the additional problem of relating the prices for these separated items. To overcome these difficulties some surveyors offer tenderers the bill in either traditional or elemental formats, computer sortation making this particularly easy. Other formats of bills of quantities are described in Chapter 24.

GENERAL PRINCIPLES

It must be remembered that in most cases the bill is a contract document and therefore the descriptions must be absolutely clear and usually without abbreviations apart from those in common use. The biller must have sufficient knowledge of construction in order to understand the descriptions and if any appear to be vague or ambiguous the taker–off should be consulted as to the exact meaning and an amendment made if necessary. The draft bill should be written on one side of the paper only, the back being used for items which may have to be inserted at a later stage, such items being clearly referenced to the exact position into which they are to be typed. After each item is written into the draft bill the corresponding item on the abstract or cut and shuffle slip should be run through with a diagonal line. It is then clear which items have been billed and which remain to be dealt with.

ORDER OF ITEMS IN THE BILL

The order of items in the bill is a matter for personal preference but the following general order is suggested:

(1) Work sections as contained in the SMM although locational sections such as substructure or external works may be required.
(2) (a) Sub–divisions of work sections as contained in the SMM
  (b) Sub–divisions as required by the SMM such as external paintwork
  (c) Sub–divisions of different types of materials such as different mixes of concrete or different types of brick.
(3) Within each sub–division in 2 above the order of cubic, square, linear and enumerated items.
(4) Labour only items should precede labour and material items within the sub–divisions in 3 above.
(5) Items within each sub–division in 3 and 4 above are placed in order of value, cheapest first.
(6) Preambles and PC and provisional sums usually form a separate bill although in certain circumstances they may be contained in the appropriate work section.

FORMAT OF THE BILL

The bill for each work section should be commenced on a new sheet, the ruling of the paper and a typical heading being shown below:

BILL 4 SUPERSTRUCTURE

MASONRY

				£	
	BRICK WALLING Common brickwork in cement, lime and sand mortar (1:1:6)				
A	Walls, half brick, vertical	97 m^2			
B	Chimney stacks, two bricks, vertical	3 m^2			

Use of columns: The first (left-hand) column is for item number or letter references and binding margin. Sometimes an additional vertical line is ruled to separate the binding margin from the reference column. The main wide column is for headings, sub-headings and descriptions. Frequently, as illustrated above, an unruled portion is left at the top of each page for main headings such as work sections. The next column contains the quantity of the item and the unit of measurement. Some surveyors prefer to enter the unit of measurement first so that the quantity is adjacent to the rate in the next column to facilitate the extension of the cash sum, others prefer to see the measurement figure separated from the cash figure so that there can be no confusion between the two. The last three columns are left blank for use by the estimator who enters the rate for the item per unit of measurement and the cash extension. Sometimes, however, a cash amount is entered by the surveyor in the last two columns, for example when entering p.c. and provisional sums.

REFERENCING ITEMS

It is essential that every item in the bill can be referred to and found easily. Some surveyors number each item consecutively through the whole bill whilst others prefer to use the page number followed by a letter entered alphabetically against each item on the page. The letters 'I' and 'O' are usually omitted to avoid any possible confusion with figures. If the number of items on a page exceed the number of letters in the alphabet then the lettering continues at this stage by using 'AA', 'AB', 'AC', etc. The latter method has the disadvantage that it cannot be done finally until the bill is typed as the draft will have a different number of items per page. On the other hand serial numbering has the disadvantage that items either inserted or deleted at the last moment disrupt the numbering sequence although the addition of a letter to the number of an inserted item overcomes this problem or an item deleted may have 'not used' entered against it.

UNITS OF MEASUREMENT

Great care must be taken, particularly when there is a change, to enter the correct unit of measurement. For example, an item measured as a cubic item and indicated as a superficial item in the bill may result in a considerable difference in price. Even if the error is detected and queried by a tenderer it may involve the issue of an addendum to the bill which should be avoided if possible. To assist in avoiding mistakes, particularly when items are inserted, it is preferable to enter the unit of measurement against each item rather than using two dots to indicate repetition.

DITTO (OR DO)

Bill descriptions may be shortened by the use of the word 'ditto' to indicate the repetition of some words which are the same as those in the item immediately before. It is essential to make it quite clear to the estimator what is intended to be covered by ditto. One ditto may be sufficient to cover the repetition of a short description, whereas a longer description may require two or three. A good rule is to assume that the word ditto repeats everything in the preceding description until it is superseded by a change in words or figures. For example:

12 mm ceiling plasterwork in two coats to concrete
12 mm ditto to isolated concrete beams

Here it is where the word 'to' occurs in the first description that the new wording commences in the second description.

Sometimes it is advisable to repeat key words to make the meaning absolutely clear. For example:

excavate trench for drain not exceeding 200 mm diameter and average depth not exceeding 750 mm
ditto average depth not exceeding 1000 mm
ditto average depth not exceeding 1250 mm

With the words 'not exceeding' used twice in the first description it is safer to include 'average depth' in the second and third description. These could be followed by:

ditto drain 225 mm diameter and average depth not exceeding 1000 mm

Which indicates a change in the diameter and average depth. Where another trench for the same diameter pipe is required but of different depth it would be described as follows:

ditto do average depth not exceeding 1250 mm

This indicates that a 'ditto' in a description following another which already contains ditto only repeats the words covered by the first ditto. Notice that the second time ditto is used in a description it is abbreviated to 'do'.

The word ditto should not be used in the first item on a page, the description being written in full or shortened by referring back to the description on the previous page with a specific reference e.g. 'all as item 265K'. Finally if there is the slightest risk that the description could be misread, then the description should be given in full.

ORDER OF SIZES

Sizes or dimensions in descriptions should always be in the order length, width and height. Sometimes the width or front to back dimension of, say, a cupboard is referred to as its depth. If there is likely to be any doubt the dimensions should be identified. For example:

sink base unit 100 mm long × 600 mm wide × 900 mm high overall ...

USE OF HEADINGS

A generous use of headings will not only help estimators to find their way about the bill but may also reduce the length of descriptions. Headings generally fall into one of four categories:

(a) work section headings such as Masonry
(b) sub-section headings such as Brick walling
(c) headings which partly describe a group of items to follow such as Common brickwork in cement mortar (1:3)
(d) sub-divisions required by the SMM such as 'Cold water'

Work section headings are often repeated in the top right hand corner of each page and other headings should be repeated when a new page is started. This enables it to be seen at once what is being dealt with, thus avoiding the necessity to turn back pages. A new heading usually terminates a previous heading and where a new heading is not used or if there is likely to be any doubt, 'end of ...' should be entered in the appropriate place in the bill.

WRITING SHORT

It is often convenient for pricing purposes to keep items together in the bill which would, following the normal rules, be separated. For example it is probably easier for an estimator to price the fittings on a gutter whilst dealing with the particular gutter although it means that enumerated items have to be dealt with in the middle of linear items. The method of entering these items is known as 'writing short' which avoids breaking completely the sequence of linear items as follows:

D	112 mm UPVC straight half round gutter fixed to timber with standard brackets	125 m			
E	Extra for stop end	9 nr			
F	Extra for angle	7 nr			
G					
H	125 mm Ditto	34 m			
J	Extra for stop end	2 nr			
K	Extra for angle	3 nr			
L					

As the example shows, written short items are inset so that the main items stand out. For the purpose of the next linear item the 'written short' items can be ignored; so that the 'ditto' in the second main item in the example above refers back to the first item. In the written short items the words 'extra for' have been repeated rather than using ditto, thus avoiding any possible confusion with the use of ditto in the main item. Items which are written short automatically refer to the main item under which they are written without the need to refer back. The SMM may require items which are often written short to be described as 'extra over'; this means that the main item has been measured over the written short item. In other words the estimator is to allow for the extra cost of the item only as it displaces the main item. Extra over is sometimes omitted from the description when the item is written short, but it would be safer to mention that this has been done in a preamble. Writing short is usually confined to enumerated items although some surveyors write short linear items which relate closely to superficial ones. For example a linear item of rounded edge tiles may be written short to the wall tiling. There must, of course, be a close relationship between the two items and it must be considered more convenient for pricing when presented in this way.

UNIT OF BILLING

The general unit of billing for other than enumerated items is the metre. The dimensions and their collections having been taken to two places of decimals, the bill entries are rounded either up or down to the nearest metre. The main exceptions are steel bar reinforcement and structural steel which is billed in tonnes to two places of decimals.

FRAMING OF DESCRIPTIONS

The framing of descriptions was introduced in Chapter 5 (Descriptions) and care must be taken to leave no doubt as to their meaning, particularly as the bill is a legal document. The opening phrase of a description should indicate the principal part of the item so that the estimator knows immediately what it is about.

The word 'approved' should be avoided if it leaves any doubt about the quality of the item. Its sole justification is to point out to the contractor that the architect's approval should be obtained for the article to be used where there are several alternatives all costing about the same. If some particular material is described and the words 'or other approved' are added to the description, a gambling element is at once introduced which is directly contrary to the purpose of the bill of quantities. One estimator may decide to price for a cheaper alternative in the hope that the architect will approve it, whilst another may price the material specified. The words 'or other equal and approved' are sometimes used by public authorities after the specification of a proprietary article to avoid the criticism that a monopoly situation has been created without any element of competition and giving favouritism to a particular firm.

Long-winded descriptions should also be avoided and any superfluous words omitted. There is a very difficult art to be cultivated by the surveyor in making descriptions concise and easily understood and at the same time omitting nothing which is essential to the estimator for pricing. In the following the words underlined, for example, could be left out:

Small 19 × 19 mm wrot softwood cover fillet planted on around bathroom door frame
50 × 175 mm Sawn softwood in floor joists spiked to timber wall plates

It should be noted that in the above, if fixing is not specified, it is at the discretion of the contractor. In these cases planting or spiking would normally be omitted as it is the most economical way of fixing. Furthermore the location of an item is not normally required.

Consistency in spelling is important if only to make the bill look like a workmanlike document. Often there are alternatives for spelling technical words such as lintel or lintol and cill or sill. An easy solution to this problem is to establish a 'house rule' that spelling should be as shown in the SMM. Misspelt words are another problem and this, whilst often unfairly attributed to typist's errors, can only be eradicated by careful reading over; although 'spell checkers' in word processing systems can assist, particularly if technical terms are added to the memory.

Furthermore consistency in language is important as minor differences in phraseology may cause an estimator trouble in deciding if the meaning is intended to be different. For example:

(a) raking cutting to existing Code 4 milled lead
(b) curved cutting on ditto

In this case there is no actual difference to be inferred from the word on. In some cases, however, descriptions which achieve the same finally may have different values, for example:

(a) formwork to rectangular mortice not exceeding 500 mm girth in concrete depth not exceeding 250 mm
(b) cutting mortices in in situ concrete 200 × 200 mm depth 100–200 mm

In the first example the mortice is formed by inserting formwork prior to pouring the concrete and in the second example the mortice is cut in the hardened concrete.

TOTALLING PAGES

There are two ways in which the surveyor may indicate how the cash totals on each page are to be dealt with. Firstly the total may be carried over to be added to the next page and so on to the end of the section. The foot of the page is completed as follows:

	Carried forward	£		

and the top of the following page as shown below:

	Brought forward	£		

The end of the section would be completed as follows:

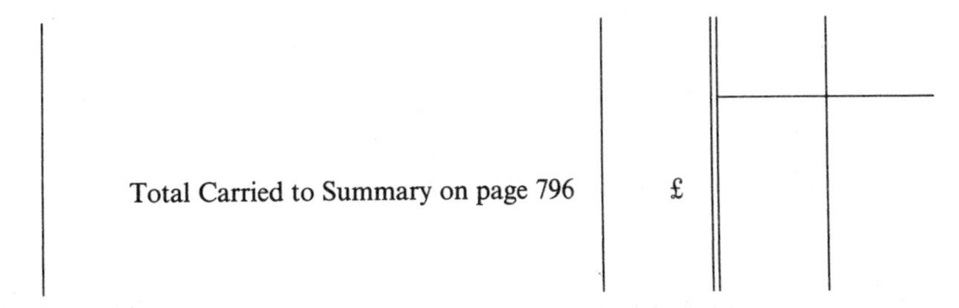

This method is only suitable when the section contains a few pages but any mathematical error made at an early stage is carried through all pages and when discovered involves several corrections and re-totalling of pages. Generally this method should be avoided.

Secondly each bill page may be totalled and each of these totals carried to a collection at the end of the section. The bottom of each page would be completed as follows:

	Carried to Collection on page 84	£		

The collection being entered on the last page of the section as follows:

			£	
	COLLECTION			
	Total from page 80			
	page 81			
	page 82			
	page 83			
	page 84			
	Total carried to Summary on page 796	£		

SUMMARY

At the end of the Bill a summary is prepared for insertion of the totals from the collections of each section. To enable the total cost of any section to be established easily it may be advisable to enter the title of each section as follows:

SUMMARY		£	
Preliminaries Total from Page Nr 10			
Specification preambles Page Nr 56			
Substructure Page Nr 76			
etc. etc.			
TOTAL CARRIED TO FORM OF TENDER	£		

Sometimes a summary is made at the end of the work sections comprising superstructure and the total of this is carried to a general summary at the end of the bill thus reducing the length of the latter. At the end of the final summary, provision may be made for the addition of items such as insurances and water for works, the price for which may be dependent on the total cost of the work. Sometimes items to allow provision for statutory payments and additional costs from the employment of workpeople, such as National Insurance, holidays with pay and guaranteed time are also included. All such items are described fully in the preliminaries and a note made against them that they are to be priced in the summary.

22 Bill preparation – preliminaries

GENERALLY

The drafting of the Preliminaries, Preambles and PC and Provisional Sums sections of the bill, which are not usually derived from the dimensions, is discussed in this chapter. These sections are often left to a late stage in the bill preparation, as reference to the draft bill of measured items (or sorted cut and shuffle slips) may be necessary. Where a full specification is supplied this will form the basis of these sections and if items have been 'run through' in the specification as they are measured then those remaining may have to be either included or, if the specification is being provided with the tender documents, referred to in these parts of the bill. It should be noted, however, that specification items which relate to method and quality of workmanship are usually excluded from the bill unless they affect cost. With the introduction of the 'Common Arrangement for Specifications and Bills of Quantities' the drafting of preliminaries and preambles is simpler and the risk of something being missed is greatly reduced.

PRELIMINARIES AND GENERAL CONDITIONS SECTION

It must be remembered that the bill of quantities has to set out all the circumstances and conditions which may affect the contractor's estimate. Items which are of a general nature and do not necessarily relate to the quantity of permanent work are set out in the preliminaries section for pricing by the estimator as method related charges divided into time related and fixed charges (SMM A, D1 and D2). A time related charge is one which is considered not to be proportional to the quantity of the item but to the length of time taken to execute the work. A fixed charge is one which is neither considered to be proportional to the quantity of work nor the time taken. The various parts of the section are discussed below.

LOCATION DRAWINGS

Three types of drawings have to accompany the bill of quantities, as follows:

(a) block plan which shall identify the site and outline of the building in relation to town plan or other wider context
(b) site plan which shall locate the position of the buildings in relation to setting out points, means of access and general layout of the site
(c) plans, sections and elevations which shall show the position occupied by the various spaces in a building and the general construction and the location of the principal elements

Each work section in the bill must be introduced with a statement giving the scope and location of the work unless it is shown on the above or additional drawings sent to tenderers or in the specification information. The SMM, at the commencement of each work section, lays down the exact nature of the information to be given.

PROJECT PARTICULARS (SMM A10-13)

This section 'sets the scene' of the proposed contract and full particulars of the project including the location; details of the employer and consultants, together with a list of drawings, are given. A full description of the site and details and dimensions of the new buildings and how they are to be constructed must be set out either by the provision of drawn information or measurement descriptions (SMM A.M1). Details have to be given about existing buildings if their presence is likely to affect the cost of the new work, for instance a mediaeval listed building or a multi–storey car park with continuous traffic problems.

CONTRACT PARTICULARS (SMM A20)

Particulars of the contract have to be given but if a standard published form of contract is being used then the conditions need not be repeated in full in the bill provided that the form of contract is named and the numbers and titles of the clauses are listed. The decision as to which clauses require pricing is left to the estimator. If the conditions of contract are not standard then the conditions must be included in full in the bill or alternatively presented as a separate volume with the tender documents and listed in the bill for pricing as above.

MODIFICATION OF CONTRACT CONDITIONS

Special clauses may sometimes be added to the conditions of a standard form of contract and these together with any modifications to clauses should be detailed in the bill. Such amendments must also be made to the actual contract which is to be signed by the employer and successful contractor, as it is not advisable to rely on a clause in the bill alone even though it may be one of the contract documents. Clauses 2.2.1 and 14.1 of the JCT form emphasise that the bills of quantities are concerned only with the quality and quantity of work, and the former clause makes it clear that the bill cannot override the contract.

EMPLOYER'S REQUIREMENTS AND CONTRACTOR'S GENERAL COST ITEMS (SMM A30-44)

Broadly speaking this part of the preliminaries section contains two types of items, firstly, those which have a separate identifiable cost such as insurances, site facilities and various fees payable and secondly, those costs which arise from a particular method of carrying out the work. The latter would include fixed costs for items such as providing the plant and bringing it to and removing it from the site, and time related costs for items such as maintaining the plant on site or for providing supervision for the works. Employers' requirements, which have to be described as fixed and time related, include the provision of site offices, fences and name boards, insurances and other fees and charges. A list of suggested items to be included under this heading is given in the SMM (A36). Limitations such as working hours and sequence of work required by the employer have also to be given. Contractor's general cost items, which again have to be described as fixed or time related, include for example supervisory staff, site accommodation, facilities such as water, power and lighting, plant and temporary works. Also included is an item for general attendance on sub–contractors which is dealt with more fully under PC sums below.

PC SUMS

These have already been discussed briefly in Chapter 19. The extent to which these are defined in the form of contract should be examined carefully as some supplementary definition may be necessary. Separate sections (A51-52) are included in the SMM relating to PC sums for nominated sub–contractors and suppliers. In both cases the contractor must be given the opportunity to add for profit to each sum. There must be no doubt as to what cash discount for prompt payment the contractor can expect in relation to PC items in the bill. If a 5 per cent cash discount is allowed then the contractor may decide to add little or no profit, but if there is no discount at all then no doubt an addition will be made to cover profit. It will be seen, therefore, that tenders could be affected substantially by any doubt as to the discount. The conditions of contract in JCT 80 state that PC sums for work to be carried out on site by 'nominated sub–contractors' referred to in clause 35 are to include a cash discount of 2½ per cent. and PC sums for goods to be supplied by 'nominated suppliers' referred to in clause 36, and fixed by the contractor, are to include a cash discount of 5 per cent. Each time an architect requests quotations from firms for PC items they must be careful to state which percentage discount is to be applied and, of course, the estimator when pricing has to think which discount will be allowed and price accordingly. The type of clause, which was seen in some contracts, saying that the contractor may retain any cash discount allowed not exceeding 2½ per cent. (or 5 per cent.) is questionable as it introduces an unnecessary gambling element. If net quotations are received to cover PC items, then the quantity surveyor must be careful to make the necessary adjustment to the sums before entering them in the bill. To adjust a net quotation to allow for 2½ per cent. discount one thirty-ninth has to be added and to adjust a net quotation to allow for 5 per cent. discount, one nineteenth has to be added. If the quotation provides for a 2½ per cent. discount and this should be 5 per cent then one thirty-eighth has to be added. This demonstrates how important it is to distinguish between PC sums for nominated suppliers and those for nominated sub–contractors. It is comon practice nowadays to have a separate section in the bill for PC and provisional sums with each type being listed under headings. One advantage of providing a separate section in the bill is that it facilitates adjustment in the final account and provides for easy identification of the various types of items. On the other hand the true cost of each work section is not obtained simply by taking the total of that section from the priced bill but would require the appropriate PC sums to be added. Carriage to site of the goods should always be included in the PC sum for nominated suppliers as the estimator cannot be expected to price for transport from an unknown source. This shows that the quantity surveyor has to be careful to read the 'small print' in a quotation to ascertain if anything has been excluded which should be drawn to the tenderer's attention in the bill or rectified by obtaining a revised quotation.

Work covered by PC sums for nominated sub–contractors must be described as fully as possible in order that the contractor can price preliminary items taking this work into account. General attendance, which basically provides for the use by the sub–contractor of facilities which have already been provided by the main contractor, is defined in the SMM (A42.C3). General attendance is included as an item in the preliminaries. Special attendance (SMM A51.1.3), for example, special scaffolding beyond the main scaffolding already in place, should whenever possible be described fully, otherwise it has to be covered by a provisional sum in the bill.

PROVISIONAL SUMS (SMM A54)

These are included for work for which there is insufficient information available for proper description. Provisional sums may also be included to cover possible expenditure on items which may be required but for which there is no information available at tender stage. Following this the SMM states that provisional sums may either be for undefined or defined work, i.e. work which can be described fairly fully. In respect of the latter the contractor is expected to have taken the work into account when pricing other sections, such as indirect costs in preliminaries. The effect of this is that when the actual sum expended is ascertained for the final account no adjustment to other prices in the bill is made. On the other hand if the work is undefined then other prices such as, for example, plant items may have to be adjusted when the work is valued for the final account.

Works which have to be carried out by local authorities or statutory undertakings should be included as provisional sums (SMM A53), as such bodies are not usually permitted to allow any cash discounts. Normally, however, they do allow a cash discount when quoting for ordinary work and this would be included in the bill as a PC item, for example an Electricity Board quoting for an electrical installation as opposed to connection to the mains. A contingency sum is usually included in the bill as a provisional sum. This sum covers unforeseen minor extras, the cost of which is sometimes difficult to explain to the building owner. The amount is deducted from the contract sum in the final account. A further provisional sum should be included to cover the cost of telephone calls made on behalf of the client.

A further subdivision of provisional sums is necessary when the JCT Intermediate Contract 1984 (IFC 84) is used. Under this contract there are no nominated sub–contractors or suppliers; they became domestic sub–contractors and suppliers named by the architect. The contractors pricing the bill may be provided with the chosen sub–contractors' tenders and invited to include the sums, together with any additions the main contractor wishes to make, in the main tender. Alternatively the main contractors are invited to include provisional sums to be set against sub–contractors' tenders in due course. When this is the case these provisional sums should be identified separately.

SPECIFICATION PREAMBLES

Preambles can be defined as general descriptions of workmanship and materials relating to the work sections which may affect price but are probably better excluded from descriptions of items. For example it may be necessary to describe cement used in making mortar for brickwork to be to BS 12 but it would not be practicable for this to be included in every description of brickwork involving the use of mortar. Commonly, preamble items form a separate section in the bill away from the work sections although many surveyors prefer to insert them at the commencement of each appropriate work section. Preambles may also include methods of measurement adopted for items which are not included in or are contrary to the SMM. If a full specification has been prepared by the architect using the common arrangement and has a similar layout to the bill then this can be referred to instead of writing preambles in full, but it is important to ensure that they are compatible and can be cross–referenced readily. Of course it is necessary to supply a copy of the specification with the tender documents and to make it a contract document.

DAYWORK (SMM A55)

Daywork is work for which the contractor is paid on the basis of the cost of materials, labour and plant plus an agreed percentage for overheads and profit. Payment in this way is usually reserved for items which cannot be measured and priced in the normal way. Daywork payments may arise in contract variations for items such as breaking up unexpected obstructions in the excavations or for adjustment of Provisional Sums. To enable a percentage rate for overheads and profit to be established at tender stage it is necessary to include items in the bill and, in order to obtain competitive rates, these items should be included in such a way that they affect the total amount of the tender.

One method of achieving this is to include a number of hours for both labourers and craftsmen which the estimator prices at an hourly rate and extends into the cash columns of the bill. Some surveyors split these hours into various crafts and, of course, it must be made clear that these rates must include for all charges in connection with the employment of labour, plus profit and overheads etc. Sums can also be included in the bill for plant and materials and the tenderer is invited to add a percentage which represents what will be added to the net cost of these items for incidental costs, overheads and profit, should the need for daywork arise.

COVER, ENDORSEMENT AND INDEX

The front cover sheet of the bill must be drafted to suit the circumstances; it should have the title and location of the work, the date and the architect's and surveyor's names and addresses. If the volume is a portion only of a larger bill, then the contents of the particular part should be defined clearly. An index on a separate sheet bound inside the front cover is useful, if not essential, especially in the case of a large bill or where the bills for a number of different estimates are bound together. Some surveyors repeat the information from the front cover on the back cover as an endorsement which is traditionally restricted to half of the cover divided vertically in the fashion of a legal document.

BINDING

Binding the document is very much a matter of personal choice. Whilst the printed cover, possibly in two colours with the client's logo and stiff boards may look well, economics have to be taken into account and a plain duplicated cover with simple spine binding is usually sufficient for all but the most prestigious of projects.

23 Checking the bill

THE PROCESS OF CHECKING

When the bill has been written (or typed in the case of a cut and shuffle system) it must be checked very carefully against the abstract or slips. Each item as it is checked should be ticked in red on the left of the draft bill and the item in the abstract or on the slips run through in red. The various points to be looked for are:

(a) In each item:
  (1) correctness of figures;
  (2) correctness of unit of measurement, watching particularly changes from one unit to another, i.e. from cube to square, square to linear, etc. properly indicated; and
  (3) correctness of descriptions, especially any figures contained therein, making it quite clear whether these are in metres or millimetres.

(b) Generally:
  (1) sections and subsections of the bill headed properly and headings carried over where necessary;
  (2) order of the items; and
  (3) proper provision for page totals and their carrying forward or collection.

These points may appear to be fairly obvious, but it is extremely easy in the rush to finish a bill, which is not unusual, for some inaccuracy to be overlooked. A 40 mm fascia in the abstract or on the slips may be billed as 45 mm, especially if the figures are poorly written. Special care is necessary when checking items transferred or inserted from the back of the sheet to ensure that such items do not disturb the sequence of descriptions, perhaps making meaningless a ‘ditto’ or altering a ‘unit of measurement’ in the item following the insertion.

NUMBERING PAGES AND ITEMS

It is important to see that the pages of the draft bill are numbered in sequence before the checking is commenced, as a page lost which would be noticed when checking might not be spotted afterwards. If a later section has to be checked prior to an earlier one, the pages should be numbered temporarily and the section pinned or tagged together so that the pages cannot be separated or mislaid. When the bill is complete it should all be fastened together and looked through, making sure that all the pages are present and in the correct sequence and that all items are referenced either by letters or sequential numbering, the latter probably being best done at this stage. If they are not present the covers to the bill, summary and index, should be inserted if they are required.

TYPED PROOF

A word may be said, perhaps, about reading over the proof prepared by a typist, which is a further stage in checking the final document. The reading should be done from the handwritten draft as the typist may have misread a word or figure and this interpretation, if read out, may be accepted as correct by the person looking at the draft. Besides actually comparing the words and figures and ensuring that all the figures are in the correct unit, all of which are, of course, important, special attention should be given to the repetition of headings and the first items on each page to ensure that ditto has not been used. This is important as the pages will almost certainly be ended at different places in the proof from the draft. It is easy for a typist to miss out an item altogether if similar ones follow each other, particularly if the quantities are the same, and particular care should be taken when reading over these items. If the bill has been typed from cut and shuffle slips then the checking process mentioned at the start of this chapter avoids the necessity to read over the proof.

GENERAL FINAL CHECK

The whole of the dimensions and abstract or slips should be examined carefully to ensure that all calculations have been ticked and all items have been run through. Any item not dealt with should be drawn to the attention of the taker–off or other person concerned. It is a good idea to put a large coloured tick at the bottom of each sheet to indicate that it has been so examined. An opportunity should be provided for the taker-off to look through the complete bill as intentions may have been misinterpreted and errors in descriptions may stand out where they could otherwise pass unnoticed. It should be noted that taking-off is not checked unless, perhaps, a junior is measuring for the first time. Therefore it is advisable to make a check of the main quantities in the bill by making comparisons between items or by making approximate calculations possibly on the following lines:

(1) check total of cube excavation with total cube of disposal of soil;
(2) compare topsoil excavation and area of floor finishes per storey, making allowances for the external walls and projection outside the building
(3) calculate the total floor area of the building on all floors inside the external walls and compare this with the total area of floor finishings and beds
(4) make an approximate measurement of the external walls, deducting all openings, and compare with the total area of walling in the bill
(5) count the number of doors and windows and compare with the total numbers of each in the bill. Ironmongery quantities also could be compared with the totals
(6) check the number and type of sanitary appliances with those billed
(7) measure the approximate area of roof tiling or flat roofing and check with the bill
(8) make an approximate check on any items which lend themselves to this, e.g. length of copings and eaves gutters, number of cupboards or other fittings, area of external pavings, number of manholes etc.
(9) compare the painting items with the corresponding items in other sections as far as possible, e.g. take total area of all wall or ceiling plaster compared with the areas of decoration on plaster

These checks must not be expected to produce an exact comparison. Their purpose principally is to ensure that the quantities are not wildly out through some serious mistake, and if the figures are reasonably near those in the bill the checks will have served their purpose. Further items for checking, other than those mentioned above, will occur to one when looking through the particular bill. Even if this check owing to pressure of time is left until the bill has been sent out to tenderers, errors found may be notified to and taken into account by tenderers. If would not, of course, be practicable to check a large number of items in the bill as it would take too long. However, if errors are found when carrying out the checks mentioned above or if there is some doubt about the taker–off's ability, then time should be afforded to create a closer check. The bill of quantities has traditionally represented the main publicised output of the surveyor's practice and careless mistakes lead to criticism from all who handle it including, one must remember, the unsuccessful tenderers.

24 Alternative billing systems

This book has concentrated mainly on the production of traditional bills of quantities by well tried and established methods. This chapter outlines some of the alternatives which have been experimented with as well as possible developments.

STANDARD BILLS OF QUANTITIES

By this stage the reader will be aware that the art of taking–off, to a large extent, is based on developing a systematic approach, having a sound knowledge of building construction, acquiring the mathematical skill to calculate and measure dimensions and the ability to write clear descriptions. Of these, probably the latter is the most difficult to master because it has to be built up from long experience, especially in estimating and dealing with errors and claims arising from inadequate descriptions. The taker–off has to ensure that descriptions comply with the requirements of the SMM and refer properly to British Standards etc., and often has to frame them under pressure of time from complex and sometimes incomplete drawings and specifications. It is no wonder, therefore, that occasionally errors and omissions occur in descriptions and that it is not unknown for there to be diversity between different bills from the same office or even between different sections within the same bill, particularly in elemental formats. Furthermore both personality and experience influence the taker–off's approach to compiling descriptions; also brevity, verbosity and English ability all have their effect. Sometimes the differences do not matter and some would argue that it is desirable to bring personality into the bill but, as previously mentioned, it can often be very irritating for the estimator who has to look for subtle differences in descriptions.

In former years an assistant became proficient in framing descriptions by experience gained in a 'working–up' section preparing abstracts and bills. With the introduction of 'cut and shuffle' and computer systems, coupled with full time education, this form of experience has been lost and alternatives have had to be established. The experienced taker–off is a comparatively expensive member of staff and the task could be open to a wider range of employees if the framing of descriptions is standardised in some way.

There are two common ways in which standardisation can be achieved. Firstly, a standard bill can be compiled in the office including all the descriptions which are likely to be encountered in the type of work usually dealt with. If an attempt is made to include every possible permutation of description the standard bill becomes a very bulky document. Another difficulty is that often the very description for which the inexperienced taker–off requires guidance is not present. Secondly, a standard libraries can be used, either being devised in the office or by adopting one of the published forms.

STANDARD LIBRARIES

Most libraries rely on the fact that phrases within descriptions are frequently repeated whereas the complete description is often peculiar to one situation. Therefore by splitting descriptions into phrases and allocating the phrases to various levels, full descriptions can be built up by selecting phrases at each level, some being essential and others optional. Considerable saving in bulk is achieved as phrases which are frequently being used are listed once only.

The levels of phrases in a standard libraries could be as follows:

- level 1 : main work section
- level 2 : subsidiary classification
- level 3 : main specification in heading
- level 4 : description of item
- level 5 : size and number
- levels 6 and 7 : written short items

By the use of tabs and pages of differing colours reference is made much easier. By selecting obligatory and optional headings and phrases from each level, combinations are built up to complete the description. The advantages of such a system are numerous, some of which are summarised in the following list:

(a) descriptions are standardised and consistent
(b) provides an aide memoire for the taker–off
(c) billing (if still an office function) can be carried out by less trained staff
(d) bill editing is largely eliminated
(e) consistency between bills aids price comparison of items
(f) simplest possible wording, almost in note form, avoids ambiguities
(g) the use of ditto (often causing doubt) is avoided
(h) complies with the SMM
(j) exposes deficiencies in the SMM and enables qualifications to the SMM to be introduced
(k) following (j) above engenders future work on the SMM
(l) permits the use of alternative bill layouts
(m) may aid computerisation of bill preparation
(n) improves communication between the surveyor and the contractor
(p) consistency between bills from various sources aids the estimator

This list would appear to give standard libraries an overwhelming advantage but there are those who are opposed to standardisation and the resulting 'staccato' style of bill produced. There may be a danger that the taker–off will apply a readily available standard description to the item being measured even though it does not fit exactly, rather than compiling carefully a proper description commonly called a 'rogue item'. There is no doubt that if the taker–off has to keep referring to bulky documents the train of thought is interrupted and speed is reduced. Furthermore the interest, initiative and possibly the incentive of the taker–off may all be affected adversely. It is costly to provide each taker–off with a complete set of documents, added to which there is the time and effort involved in keeping each set up to date.

In order to show a comparison between a traditional bill description and one obtained from a standard libraries the following examples are given:

Traditional:
<u>Reinforced precast concrete (Grade D) as described</u>

900 × 110 × 150 mm rectangular lintel reinforced with one 12 mm diameter mild steel bar

Standardized:
PRECAST CONCRETE
Normal: mix 21 N/mm^2: 14 aggregate: vibrated; reinforcement bars BS4449. hot rolled plain mild steel
Lintels; plain cuboidal shape; 900 × 110 × 150; reinforced 1.4 kg 12 mm bar.

COMPUTERISED BILL PRODUCTION

It is beyond the scope of this book to describe the available computer systems in detail but when outlining bill production generally it is obvious that a small amount of space must be given to discussion on the main points concerning computer systems.

Generally, quantity surveyors who became involved with computer systems some twenty years ago seemed to find them somewhat disruptive and of no great financial benefit over manual systems of bill production. Furthermore, cut and shuffle systems became more efficient as firms became more familiar with them and this added resistance to the use of computer systems. However, with rapid advances in computer technology and with cheaper and more compact hardware, many firms have now adopted computer systems. Popularity has not only emerged for their use in bill production but also for assistance in carrying out most other quantity surveying functions. It is desirable that the operation of these other functions can be carried out without having to re-enter the data stored for bill production.

Most systems require the taking-off descriptions to be coded in some way but at the same time aim to cause as little disturbance to the taking-off process as possible. Frequently, however, the taker-off is required to use specially ruled paper to enable the code to be entered against the descriptions ready for computer input. Whether the coding is done by the taker-off or by coding specialist is usually a management decision but if done by the former it is probably less disruptive if descriptions are coded in batches rather than individually. Systems have been developed, however, where no paper is used and the taker-off works on a digitiser and keyboard and therefore codes the descriptions one at a time.

All systems depend on having a library of descriptions which should be easy to use, flexible and easily updated. Once the dimensions and coded descriptions have all been entered into the machine the bill production takes a matter of hours rather than days. Another advantage is that alternative bill formats may be produced readily. One important feature should be that an abstract or a proof of the bill is obtained for editing and that amendments to the stored data are easily made for the final print out. Editing is probably more important than when using manual methods of bill production as the smallest coding error will be reflected in an incorrect bill description. The final product must be presentable and easily read otherwise it will not be popular with estimators.

It is possible to measure repetitive standard details by inserting basic information into the machine, possibly from schedules. Such work as drainage, excavations and steel reinforcement is suitable for measurement by such methods.

Mention should be made at this stage of Computer Aided Architectural Design (CAAD) systems which, although at the moment are bulky and expensive, will no doubt become more popular. These systems can have the ability to produce measurements and therefore bills of quantities particularly on new projects with a high degree of repetition. Whilst, as with most professions, advances such as this will in years to come affect the work of the quantity surveyor one must remember that there is much work to be done in developing and designing the systems apart from the expertise required to understand and manage their output. Computer drafting systems to produce working drawings are no doubt likely to become more widespread and in the first instance some form of quantities automatically devised from these will probably be in the forefront of this type of development.

ALTERNATIVE FORMATS FOR BILLS OF QUANTITIES

Over the years there has been certain dissatisfaction expressed with the traditional bill of quantities format from both contractors and architects but, additionally, there appears to be a certain reluctance either to produce or accept other types of arrangement. The elemental bill, where the work is presented in the order of work sections within sub-divisions of elements, has already been mentioned. This format suits cost analysis and therefore is convenient for the quantity surveyor but some estimators find the repetition of items lengthens the pricing operation. To overcome this difficulty 'sectionalised trades bills' were introduced, which are traditional bills in which each work section is divided or sectionalised into elements. In some cases a loose–leaf arrangement is used, with each new element entered on a separate sheet, thus enabling the bill to be sorted either into work sections or elements. There is still the difficulty of repetition of items but this is in a more concentrated form. Over the last quarter century various other bill formats have been tried and some have met with limited success. Some of these alternatives are described briefly below because the experience gained in their trial may be of some value when devising new formats in the future.

OPERATIONAL BILLS

The operational bill was devised by the Building Research Establishment (BRE) in the late 1960s in an attempt to meet the criticisms that the traditional bill failed to convey all the useful information about the proposed building which is generated in the measurement process. Furthermore the method of measurement used is not considered to be the most effective way of presenting items for pricing. In the preparation of this type of bill a programme for the building work is devised and broken down into a series of short–term site operations.

Each operation is presented for pricing in a form sub–divided into labour, materials and sundries, plant and off–site work being billed under separate headings. A precedence diagram showing the relationship of the envisaged operations is included with the bill.

Whilst in theory this bill format relates the design of the building to its construction and should therefore produce more accurate estimates, experience showed a reluctance to accept it. This reluctance probably stems from the architects' resistance to produce operational drawings, the quantity surveyors' lack of expertise in preparing building programmes and the estimators' need to change radically the estimating process. Difficulty was also experienced in the valuing of work varied during the contract which often had to be done by negotiation. This bill format can obviously be described as a progressive and worthwhile innovation but it probably came at a time when the industry was too busy and traditionally resistant to change and a less revolutionary variant known as the 'Operational Format Bill' was proposed by the BRE. This format is similar to the activity bill next described.

ACTIVITY BILLS

The basis of the activity bill is similar to that of the operational bill described above, except that the operations are detailed in the bill in accordance with the rules of the SMM rather than labour and materials.

Experience has shown that this format was more acceptable to estimators but probably the degree of repetition of items, the additional time taken in bill preparation and the difficulty in preparing a precedence diagram at pre–tender stage contributed to the lack of its widespread acceptance. In the later development this type of bill was renamed by the BRE as a Bill of Quantities (Operational Format).

ANNOTATED BILL

The annotated bill may be described as a traditional work section bill in which each item is located by reference to its position in the building. This sometimes results in the division of quantities of an item for each location and sometimes involves the insertion of relevant specification clauses. The annotation is usually presented on the back of the opposite page to the measured items which facilitates immediate reference. In practice these bills were highly thought of by contractors as they were far more useful during the construction process and for valuing work for interim payments. However, the increase in size of the bill, the longer preparation time and some reluctance to use the benefits to advantage caused this type of bill to fall into disuse.

LOCATION BILLS

The location bill may be regarded as a combination of activity and annotated bills. This format was recommended by the RICS Standardisation of Format Working Party in 1965 with the view that information generated in taking–off should be published for the benefit of the contractor. These bills contain door, window and finishings schedules, guidance on phasing the work and the location of individual items.

CONSOLIDATED SCHEDULE BILLS

The consolidated schedule is a form of bill presentation developed by the Scottish Special Housing Association for repetitive work, such as housing. Traditionally a self–contained sub–bill is prepared for each house type, priced separately and timesed by the number of houses in the summary. The majority of items tend to occur with identical descriptions in each sub–bill and under the consolidated system the quantities for identical items are collected together into totals resulting in a smaller bill. This provides consistency in pricing and a saving in time for estimating.

The estimator is informed of the number of different house types in the preliminaries or tender documents. The bill of the successful contractor is analysed and resorted by computer, giving the cost of each work section for each house type enabling a break down for stage payments. In certain circumstances it is suggested that when the number of sub–bills does not exceed five a 'broadside' presentation may be used. This presentation has the quantities for like items for each house type arranged laterally for pricing simultaneously.

OTHER BILL TYPES

Other titles for bills which have a special purpose rather than a special format may be encountered and are mentioned here for reference purposes.

MASTER BILLS

These are prepared, possibly on a notional basis, for obtaining tenders for serial contracts where the work is phased over a period. The master bill forms the basis for the bills prepared for later stages of the work.

REDUCTION BILL

This is a special bill prepared when the tender figure is too high and a reduction in price is obtained by altering the work in some way. The bill may contain both omissions and additions to the original.

ADDENDA BILLS

These contain details of additional work which is required to the original design, decided upon after completion of the original bill.

SPECIALIST BILLS

These may be required to obtain tenders for specialist work, e.g. electrical installation, which is to form sub–contract work. These bills should contain the full preliminaries section of the main bill and be accompanied by the necessary drawings.

BILL OF APPROXIMATE QUANTITIES

This bill, also known as a 'provisional bill', is used when there is insufficient information available to prepare an accurate bill of quantities. Whilst the quantities are approximate the descriptions should be correct. The bill is used to obtain rates for items from tenderers and as the production information becomes available a 'substitution bill' is prepared and priced using the approximate bill as a basis.

NOMINATED BILL

In some negotiated contracts the contractor is requested to submit a bill for another contract which the firm has priced recently in competition. If agreed this bill is nominated as the basis of pricing the measured work for the negotiated contract.

SCHEDULE OF PRICES OR RATES

For smaller contracts without a bill and where the specification and drawings are the contract documents a schedule of descriptions, without quantities, can be prepared. The contractor is asked to enter rates against the items which are then used for valuing variations to the design. The schedule seldom attempts to be fully comprehensive and the rates are often unreliable. This method has also been used for contractor selection when drawn information is so scanty that even a bill of approximate quantities is impracticable. The prices in this form are difficult to check as an overall total is not obtained for comparison.

The Department of Environment Schedule is a form of price list, carefully drawn up to contain items likely to be encountered in Government work. The tenderers compete by quoting a percentage addition or deduction from the rates published in the schedule. The work carried out is measured and the bill is priced from the schedule and adjusted by the quoted percentage.

GENERALLY

The increased use of computers should enable alternative bill formats to be produced more easily and there is no doubt that there are several improvements that could be made to the ways in which work is presented for estimating purposes. Tendering by negotiation rather than competition usually allows experimentation in the use of different bill formats without causing too much disruption. It must be remembered, however, that familiarity with a well established and accepted format can inhibit people's thinking and judgement when considering a new type of bill.

Appendix 1 Abbreviations

Whilst the use of abbreviations is limited when employing certain taking–off systems, there are occasions when they may be used to save time and space and therefore some suggestions are listed below. Contractions of words are made frequently by the omission of letters and their meaning is obvious, particularly when read in context. Abbreviations in isolation may have two meanings, but again when looked at in context are usually understood. Care must be taken, however, not to use abbreviations such as 'f e' which may, when written short, mean fair end or fitted end. Generally speaking abbreviations are not used in bills of quantities except for those generally recognised such as BS and metric designations.

a b	air brick
	as before
a d	as described
a f	angle fillet
AI	Architect's Instruction
a w p	as the work proceeds
add	additional
adj	adjacent
agg	aggregate
al	aluminium
appd	approved
achve	architrave
art	artificial
asp	asphalt
av	average
B	brick
b & j	bed and joint
b & p	bed and point
b f	before fixing
b i	built in
b i g	back inlet gully
b m	birdsmouth
b m a	bronze metal antique
b m s	both measured separately
b n	bullnose
b o e	brick on edge/end
B S	British Standard
b s	both sides
bal	baluster
bast	basement
bdd	beaded
	bedded
	boarded
bdg	boarding
bdy	boundary
benchg	benching
bit	bitumen
bk	brick
bkt	bracket
bkwk	brickwork
blbd	blockboard
bldg	building
blwk	blockwork
btm	bottom
bwk	brickwork
c & f	cut and fit
c & p	cut and pin
c & s	cement and sand
	cups and screws
c b	common brickwork
c i	cast iron
c l	centre line
c m	cement mortar
c n	copper nailing
c p	chromium plated
c s	clear sheet
c/s	countersunk
c t & b	cut tooth and bond
cast	casement
chfd	chamfered
chy	chimney
circ	circular
cist	cistern
clg	ceiling
col	column

conc	concrete
corr	corrugated
cos	course
csk	countersunk
ct	cement
cupd	cupboard
cyl	cylinder
d h	double hung
d p c	damp proof course
d p m	damp proof membrane
d s	double seal
ddt	deduct
dia	diameter
div	divided
do	ditto
dp	deep
dr	door
drg	drawing
e g	eaves gutter
e m	elsewhere measured
e m l	expanded metal lathing
e o	extra over
e p	emulsion paint
e w m	elsewhere measured
emul	emulsion
exc	excavate
excdg	exceeding
ext	external
exstg	existing
f a i	fresh air inlet
f f	fair face
f f l	finished floor level
f l & b	framed ledged & braced
f o	fix only
fcg	facing
fclay	fireclay
fdn	foundation
fin	finished
fittg	fitting
fl	flush
flg	flooring
fr	frame
frg	framing
ftd	fitted
furn	furniture
fwk	formwork
fxg	fixing
g i	galvanised iron
g l	ground level
g m	gauged mortar
g s	general surfaces
g w i	galvanised wrought iron
g w l	ground water level
galv	galvanised
glzg	glazing
grano	granolithic
grd	ground
grtd	grouted
gth	girth
gv	groove
gyp	gypsum
h & c	hat and coat
h b s	herring bone strutting
h b w	half brick wall
h c	hardcore
h j	heading joint
h n & w	head nut and washer
h p	high pressure
h r	half round
	hand rail
	heat resisting
h t	hollow tile
h w	hollow wall
	hot water
	hardwood
hdb	hardboard
hdg	heading
hi	high
hor	horizontal
hsd	housed
ht	height
hth	hearth
hwd	hardwood
i c	inspection chamber
i l	invert level
inc	including
inst	installation
int	internal
inv	invert
irreg	irregular
jst	joist
jt	joint
jtd	jointed
K p s	Knot prime and stop
l	linear
l & b	ledged and braced
l & c	level and compact
l & m	labour and material
l l	low level
l m	lime mortar
l p	large pipe
	low pressure
la	large
lab	labour
layg	laying
lev	level
lin	linear
lng	lining
lt	light
m c	metal casement
m e	match existing
m g	make good
m h	manhole
m l	mortice lock
m n	measured net
m s	measured separately
	meeting stile
	mild steel
mis	mitres
mo	moulded
mort	mortice
msd	measured
n & f	notch and fit
n e	not exceeding

n t s	not to scale
n w	narrow widths
necy	necessary
nld	nailed
nom	nominal
nr	number
nsg	nosing
ø	diameter
o/a	overall
o e	one edge
	other edge/end
	one edge/end
o s	one side
	one stile
o w	old wall
obs	obscured
opg	opening
p & s	plugged and screwed
par	planed all round
p b	plasterboard
	plinth block
p c	prime cost
	precast concrete
P ct	Portland cement
p d	pull down
p e	plain edged
p m	purpose made
p p	polished plate
P s	Portland stone
p s	pressed steel
pan	panel
patt	pattern
pbd	plasterboard
pl	plate
plast	plaster
pol	polished
pr	pair
prep	prepare
proj	projection
prov	provide
	provisional
pt	part
	paint
	point
ptg	pointing
ptn	partition
pvg	paving
q t	quarry tile
r c	reinforced concrete
	raking cutting
	rough cut
r e	rodding eye
	rounded edge
r l	reduced level
r m e	returned mitred end
r o j	rake out joints
r r e	returned rounded end
r w h	rainwater head
r w p	rainwater pipe
rad	radius
rdd	rounded
reb	rebated
reinf	reinforced
rem	remove
ret	returned
rl	rail
ro	rough
s & f	supply and fix
s & l	spread and level
s & r	splayed and rounded
s a a	satin anodised aluminium
s d	screw down
s e	stopped end
s f	sand faced
	stepped flashing
s g	salt glazed
s h	side hung
s l	short length
s n	swan neck
s p	small pipe
s q	small quantities
s s	stainless steel
s v	stop valve
	sluice valve
s v p	soil and vent pipe
s w	softwood
san	sanitary
sk	sunk
sktg	skirting
sm	small
sn	sawn
soff	soffite
spec	specification
splyd	splayed
spr	spread
sq	square
str	straight
surd	surround
surf	surface
susp	suspended
t	tee
t & g	tongued and grooved
t & r	tread and riser
t g v	tongued grooved and v-jointed
tankg	tanking
tapd	tapered
temp	temporary
tgd	tongued
th	thick
thro	throated
tiltg	tilting
tlg	tiling
tog	together
tr	trench
trow	trowelled
u/c	undercoat
u/grd	underground
U/P	underpinning
V O	variation order
v p	vent pipe
ve	verge
veg	vegetable
vent	ventilation
vert	vertical
vit	vitreous
	vitrified

w & p	wedge and pin
w/s	working space
w b p	weather and boil proof
w g	white glazed
w i	wrought iron
w p	waste pipe
	waterproof
	wax polish
wdw	window
wi	with
	wide
wl	wall
wt	weight
	wrot
wthd	weathered
x grain	cross grain
x tgd	cross tongued
xtg	existing
xtl	external
Y s	York stone
③	three (oils) coats gloss paint

Appendix 2
Schedules prepared to assist with taking-off

ADVANTAGES

1 Familiarise the taker-off with extent and variety of work.
2 Expose omissions or lack of information on drawings or in specification at one early stage.
3 Enable the taking-off to be organised into a concise form.
4 Reduce the amount of working-up or sorting of slips.
5 Form a check list for taking-off (loop through items as they are measured).
6 Reduce the need to refer to drawings during taking-off.
7 Create a form of index to the taking-off.
8 Enables what has been measured to be readily ascertained.

EXAMPLES OF SCHEDULES

The headings and width of columns shown should be amended to suit the particular circumstances. More sophisticated schedules may include dimensions which could be squared and totalled for transfer to the taking-off.

INTERNAL FINISHINGS

LOCATION	CEILING		WALLS		DADO
	FINISH	DECORATION	FINISH	DECORATION	
Bathroom	9·5 Gypsum lath 5 skim	2ce emulsion	13 two ct plaster	2ce emulsion	Glzd tile 1m high

WINDOWS

Nr	LOCATION	OVERALL SIZE	TYPE	GLASS	OPENING SIZE	WALL	FINISH	
							IN	OUT
1	Lounge	1073 x 1070	Centre hung in one pane	Sealed dbl glzd unit	1075 x 1080	Half Brick	Plast Emuls	Facings

DOORS

Nr	LOCATION	TYPE	THICK-NESS	SIZE	FRAME/ LINING	DECOR-ATION	ARCHITRAVE/ COVER MLD	IRONMONGERY
A	Lnge/Hall	Flush ply fcd	40mm	726 x 2040	115 x 32 Lining	KPS & 3	75 x 25 Mo Archve b/s	Pr 75 Butts M.L. & F

DRAIN RUNS

LOCATION	c/c	LENGTHS					DEPTH		
		PIPE	EXCN	BED	HAUNCH	SURRD	TOP	BTM	AVGE
MH1-MH2	6500	675 6500 563 1238 x ½ = 619 5881	5881 2/=450 2/=300 750 5131	5131	5881 450 5431	—	675 50 150 875	825 50 150 1025	875 1025 1900 av 950

MANHOLES

Nr	DIAGRAM	INTL SIZE	DEPTH EXCN	SIZE 150 CONC BASE	SIZE 150 RC COVER	COVER & FRAME	WALLS 1B	
							HEIGHT	GIRTH
1		563x 675	675 Chan 50 Base 150 875	563 675 750 750 1313 x 1425	563 675 450 450 1013 x 1125	457 x 457 CI BS 497 Pt 1 Grade C	675+50=725 -150 575	563 675 2/1238 =2476 4/= 900 3376

CORNICE	SKIRTING	FLOOR		REMARKS
		BED	FINISH	
250mm gth moulded	25x100 rdd s w Kps & 3	20mm c&s	225 x 225 x 3 Thermoplastic	Area ne $4m^2$

LINTOL	SILL		IRONMONGERY	REMARKS
	INSIDE	OUTSIDE		
103 x 150 P conc one 16mm bar	—	Timber	With window	—

GLASS	OPENING SIZE	WALL	FINISH		LINTOL	FLOOR/ SKTG	REMARKS
			INSIDE	OUTSIDE			
—	790 x 2072	75 Blk	2ct plaster 2ce em	2ct plaster 2ce em	75 x 100 P conc 16mm bar	Meas net	—

PIPE			GULLIES	REMARKS
TYPE	SIZE mm	BENDS		
Vit Clayware BS	100	—	—	—

CHANNEL	BRANCH BENDS	B I ENDS	REMARKS
100mm Str 675 long	Two 100mm	Four 100mm	

Appendix 3
Mathematical formulae

FORMULAE FOR AREAS (A) OF PLANE FIGURES

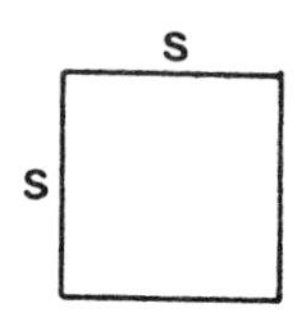

SQUARE $A = S \times S$

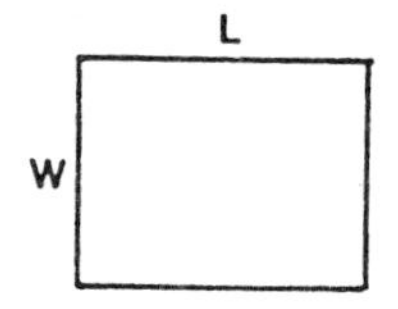

RECTANGLE $A = L \times W$

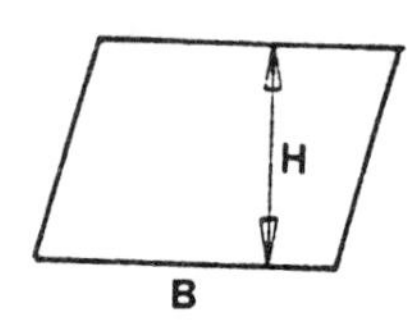

PARALLELOGRAM $A = B \times H$

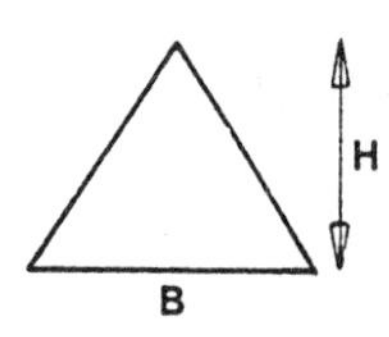

TRIANGLE $A = \frac{B \times H}{2}$

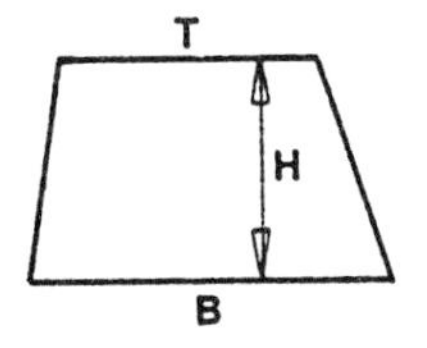

TRAPEZOID $A = \frac{(B + T) \times H}{2}$

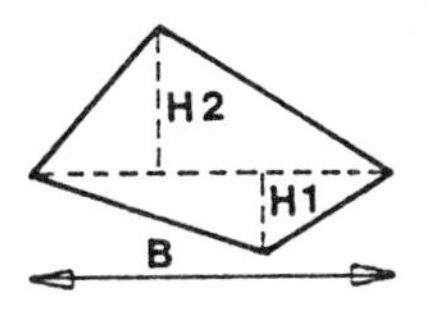

TRAPEZIUM $A = \frac{B \times H1 + B \times H2}{2}$

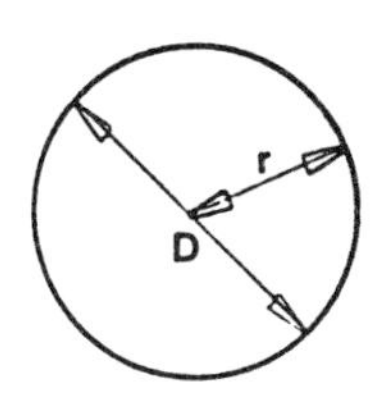

CIRCLE $A = \pi \times r \times r$

or $A = 0.7854 \times D \times D$ (Note $\frac{\pi}{4} = 0.7854$)

(Circumference = $\pi \times D$ or $2\pi r$)

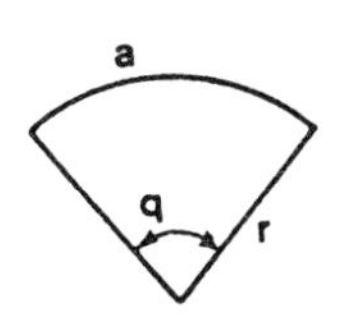

SECTOR OF CIRCLE $A = \frac{r \times a}{2}$ or $A = \frac{q}{360} \pi r^2$

(Note length of arc = angle $\frac{q}{360} \times \pi 2r$)

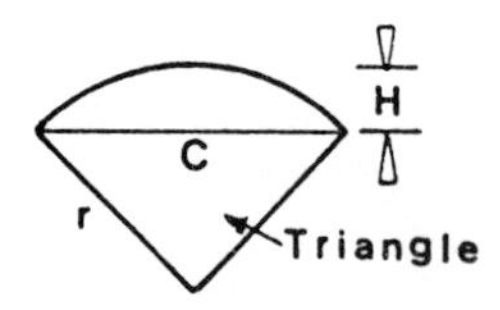

SEGMENT OF CIRCLE $A = S - T$ where S = area of sector
T = area of triangle

or approximately

$$A = \frac{1}{2} \times \left(\frac{H \times H \times H}{C}\right) + \left(\frac{2}{3} C \times H\right)$$

where H = rise
C = chord

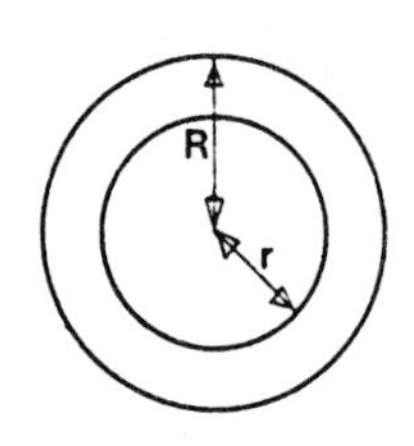

ANNULUS $A = \pi (R + r) \times (R - r)$

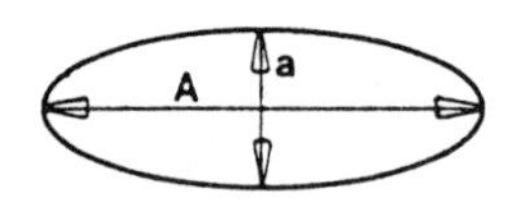

ELLIPSE $A = 0.7854 \times (A \times a)$

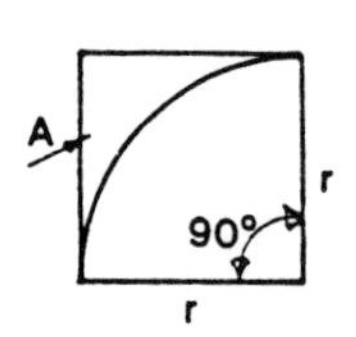

BELLMOUTH (at road junction) $A = 0.2146 \times r \times r$

REGULAR POLYGONS

Pentagon	(5 sides)	$A = S \times S \times 1.720$
Hexagon	(6 sides)	$A = S \times S \times 2.598$
Heptagon	(7 sides)	$A = S \times S \times 3.634$
Octagon	(8 sides)	$A = S \times S \times 4.828$
Nonagon	(9 sides)	$A = S \times S \times 6.182$
Decagon	(10 sides)	$A = S \times S \times 7.694$

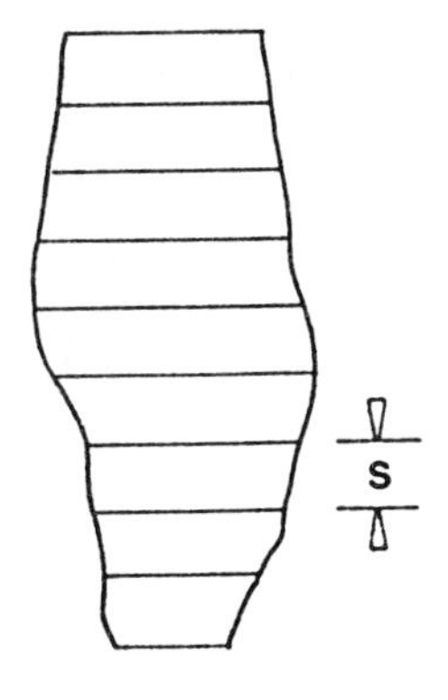

IRREGULAR FIGURES

Divide figure into trapezoids by equidistant parallel lines (ordinates or offsets)

$$A = S \times \left(\frac{P}{2} + Q\right)$$

(Where S = distance between ordinates
P = sum of first and last ordinate
Q = sum of intermediate ordinates

Or

Simpson's Rule (must be equal number of trapezoids)

$$A = \frac{S}{3} \times (P + 2 \times Z + 4 \times Y)$$

(Where Z = sum of even intermediates ordinates
Y = sum of odd intermediates ordinates)

FORMULAE FOR SURFACE AREAS (SA) AND VOLUME (V) OF SOLIDS

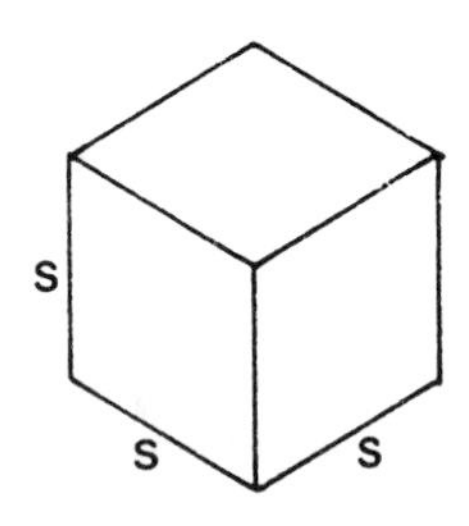

CUBE

$SA = 6 \times S \times S$
$V = S \times S \times S$

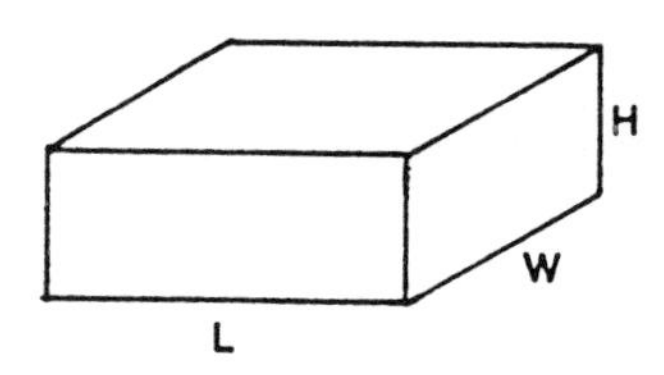

RECTANGULAR PRISM

$SA = 2(L \times W) + 2(L \times H) + 2(W \times H)$
$V = L \times W \times H$

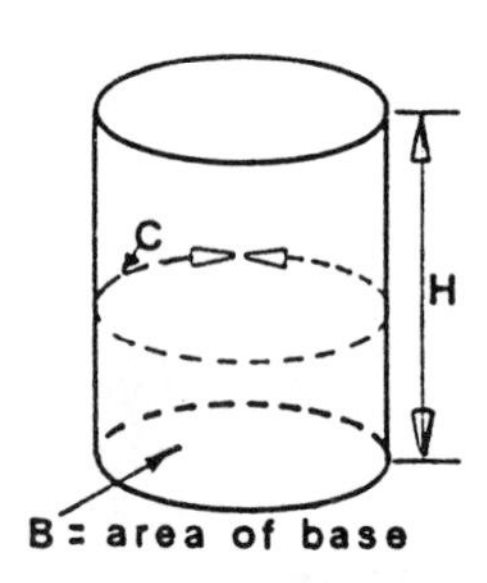

CYLINDER

$SA = (C \times H) + (2 \times B)$
$V = B \times H$

Where B = area of base $(\pi \times r \times r)$
C = circumference $(\pi \times D)$

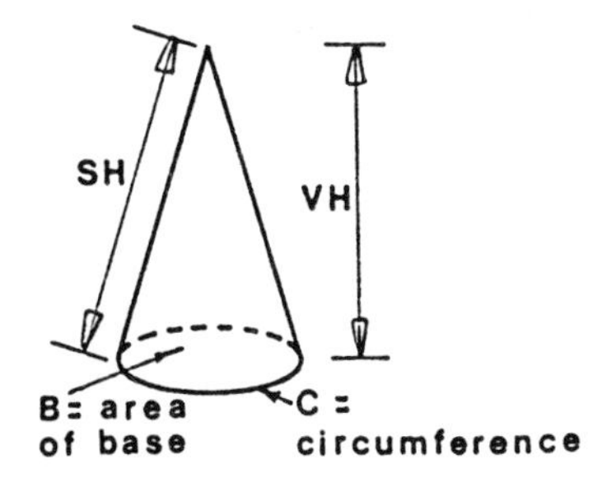

CONE

$$SA = \frac{C \times SH}{2} + B$$

$$V = \frac{B \times VH}{3}$$

Where B = area of base ($\pi \times r \times r$)
C = circumference of base

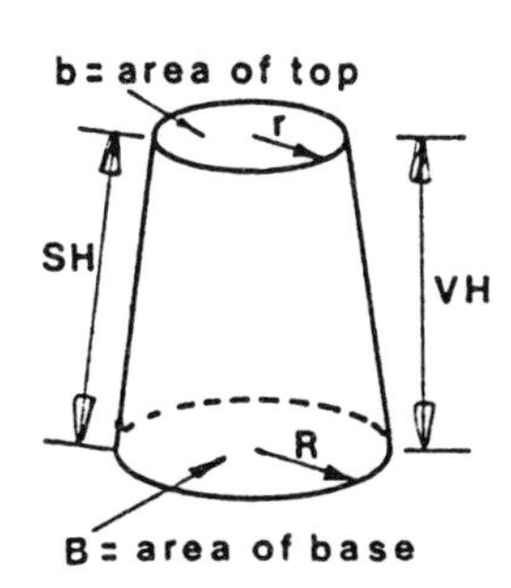

FRUSTRUM
OF CONE

$$SA = \pi \times SH \times (r + R) + b + B$$

$$V = \frac{VH}{3} (\pi r^2 + \pi R^2 + \sqrt{R \times r})$$

Where B = area of base
b = area of top
R = radius at base
r = radius at top

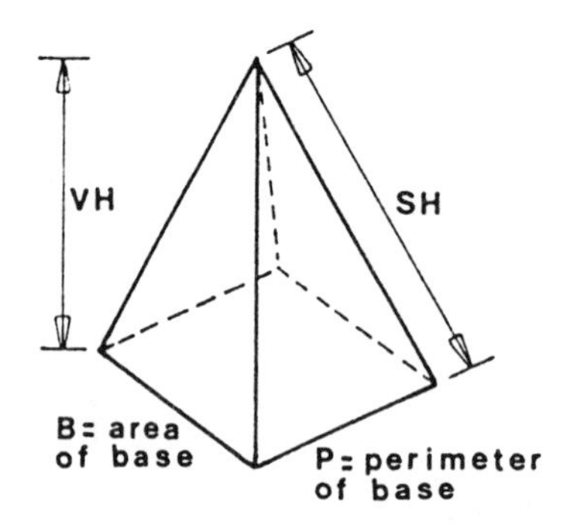

PYRAMID

$$SA = \frac{P \times SH}{2} + B \text{ (regular pyramid only)}$$

$$V = \frac{B \times VH}{3}$$

Where B = area of base
P = perimeter of base

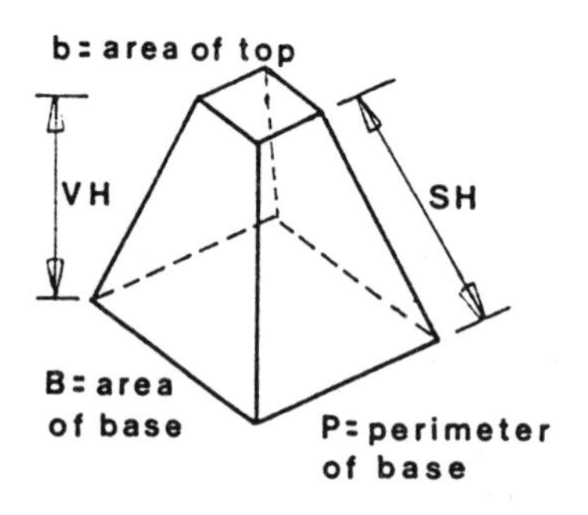

FRUSTRUM OF PYRAMID

$$SA = \frac{SH}{2}(p + P) + \sqrt{B + b} \text{ (regular figure only)}$$

$$V = \frac{(B + b + \sqrt{B \times b}) \times VH}{3}$$

Where B = area of base b = area of top
P = perimeter of base
p = perimeter of top

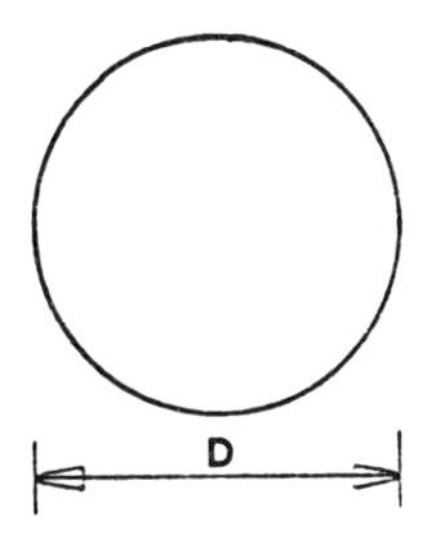

SPHERE

$SA = \pi \times D \times D$
$V = 0.5236 \times D \times D \times D$

(Note $\frac{\pi}{6} = 0.5236$)

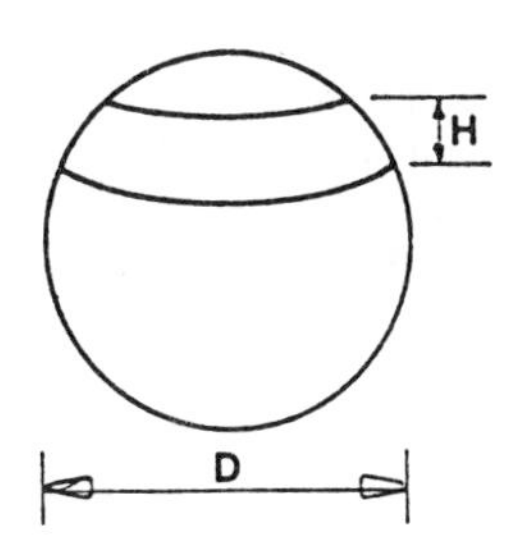

ZONE OF SPHERE

$SA = \pi \times D \times H$ (excluding base & top)

$$V = \frac{\pi \times H}{6} \times (3 \times R \times R + 3 \times r \times r + H \times H)$$

Where R = radius at base
r = radius at top

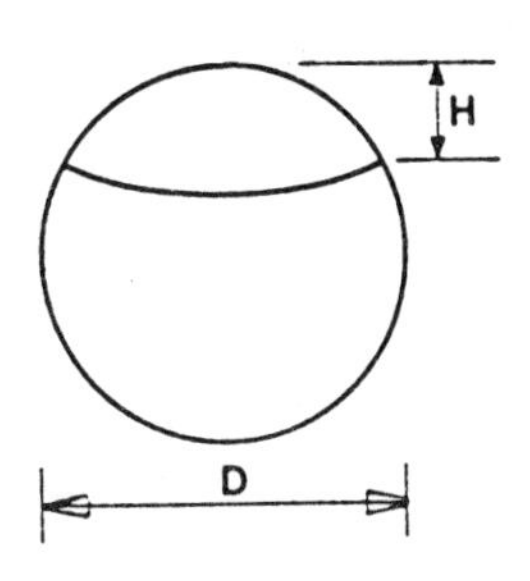

SEGMENT OF SPHERE

$SA = \pi \times D \times H$ (excluding base)

$$V = \frac{\pi \times H}{6} \times (3 \times R \times R + H \times H)$$

Where R = radius of base

Index